MACHINING

FOR
HOBBYISTS
GETTING STARTED

KARL H. MOLTRECHT

with

Fran J. Donegan, *Developmental Editor*

Foreword by
George Bulliss, *Editor, The Home Shop Machinist*

INDUSTRIAL PRESS, INC.

Industrial Press, Inc.

32 Haviland Street, Unit 2C, South Norwalk, Connecticut 06854
Phone: 212-889-6330, Toll-Free in USA: 888-528-7852, Fax: 212-545-8327
Email: info@industrialpress.com

Machining for Hobbyists: Getting Started

By Karl H. Moltrecht

ISBN Print: 978-0-8311-3510-2
ISBN ePDF: 978-0-8311-9344-7
ISBN ePUB: 978-0-8311-9345-4
ISBN eMOBI: 978-0-8311-9346-1

Sponsoring Editor: John Carleo
Developmental Editor: Fran J. Donegan
Interior Text and Cover Designer: Janet Romano-Murray

industrialpress.com
ebooks.industrialpress.com

TABLE OF CONTENTS

DETAILED TABLE OF CONTENTS

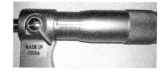

List of Tables

continued

vii

FOREWORD

By George Bulliss, Editor
The Home Shop Machinist

My day job allows me the opportunity to talk with newcomers to the machining hobby on a regular basis, many with questions about how to get started. Unlike most other hobbies, in metalworking, and in machining in particular, it can be tough to find fellow hobbyists. For those living beyond large urban areas the learning process is typically a solitary journey.

This lonely path often starts on the Internet, where the sheer bulk of information can overwhelm and confuse. Not to mention, the Internet comes with no guarantee of accuracy; so-called old wives' tales abound, and the beginner, unable to sort fact from myth, can easily head down the wrong, frustrating path.

Fortunately for those jumping into this hobby, there is a long list of quality books that can help. However, this is not without its pitfalls. For a hobby that dates back to the beginnings of the Industrial Revolution, numerous titles have been published, making the choice extremely tough.

So what does the beginner need? First, you must get a handle on the basics and make sense of common terms and techniques. Without knowing the lingo and the various tools and equipment used in machining, learning the ropes will be difficult at best.

Mastering terms and techniques is only part of the story; sooner or later one must turn on a machine and cut some metal. It's at this point that beginners discover machining metal requires knowledge of cutting parameters if they hope to avoid damaging tools and destroying workpieces.

For anyone with woodworking experience, the fussy nature of cutting metal may come as a bit of a surprise. Drilling a hole in wood is straightforward: select the drill, turn the drill press on, and run the drill through the board on your mark, with acceptable results pretty much certain.

For the machinist it's not that easy, even for something as simple as making a hole. Marking your location accurately enough for most machined components will take more than just a tape measure and a pencil. Picking the right sized drill is easy enough, but will it actually drill the correct size hole – or even make a hole? With the right cutter geometry and drill speed, making a hole in metal is an easy task, but it quickly gets expensive when you try to guess!

When I first heard of this book I was excited by its mix of material. Finding basic machining information to answer the beginner's questions and the technical information needed to actually cut metal in one book is something of a rarity. With this book's publication I finally have an answer for that oft-asked question, "What book do I need to get started?"

Thinking about taking the plunge into machining? You'll find this book makes the perfect foundation for your shop library, and the mix of information and reference material will keep it relevant and useful for years to come.

INTRODUCTION

The heart of any machine shop is the relationship the machinist has with the tools and equipment that keep the shop humming along. That's true whether it's an industrial shop turning out parts for jet engines, or a basement or garage setup for home hobbyists.

Machining for Hobbyists: Getting Started is intended for the latter group. It examines the tools and materials home machinists use to create their own projects, reinforcing the relationship between the person and the machine. It lays the groundwork for novices and even some experienced machinists to grow in their craft. Add practice and dedication, and the inexperienced user becomes an expert.

Machining metal requires specialized tools, to which this book devotes several chapters. Chapter 2, "Measuring Tools", deals with the array of tools used by home machinists to take measurements and lay out their projects. This includes everything from plain steel rulers to the various types of calipers and micrometers. There are also tips on how to use these measuring devices.

Chapters devoted to lathes, mills, and drill presses will help the novice get started in assembling the necessary tools and equipment for their projects. These tools were chosen for this book because they come in smaller bench-top sizes that make the most sense for the home machinist. You will find descriptions of the tools and their components and tips on using the equipment.

To complement the material in this book, the editors have added six articles reprinted from *The Home Shop Machinist*, a bimonthly publication geared to machinists of all levels of experience. The articles were written and photographed by two of the magazine's regular contributors and provide expert additional information on some of the tools and processes covered in the book.

Machining for Hobbyists covers manual lathes, mills, and other equipment; it does not discuss CNC machines, which are computer controlled. CNC tools have largely taken over the industrial machining industry, but for the hobbyist, learning how to measure the work, operate the tools, and solve problems by hand is still the best way to learn the craft.

Machine Shop Overview

The ability to machine metal to produce precisely engineered parts is the driving force behind large-scale manufacturing. Fortunately, the knowledge that went into producing large industrial machines is available to home hobbyists who want to build scale models of full-size items or create their own products made from metal. This book covers the selection and use of home shop-size equipment. While there are a variety of tools and equipment to choose from, *Machining for the Hobbyist* will cover the most popular, which are the tools and machines available in bench-top or home-shop size.

Machine Tools

The general term "machine tool," includes various classes of power driven metal cutting machines. Most machine tools change the shape of a material by producing chips. Machine tools serve four main purposes:
1. They hold the work or the part to be cut.
2. They hold the tools that do the cutting.
3. They provide movement of either the work or the cutting tool.
4. They are designed to regulate the cutting speed and also the feeding movement between the tool and the work.

In the production of machine parts of various shapes and sizes, the type of machine and cutting tool used will depend upon the nature of the metal-cutting operation, the character of the work, and, possibly, other factors such as the number of parts required and the degree of accuracy to which the part must he made. The development of machine tools has been largely an evolutionary process, as they have been designed to produce parts meeting increasingly stringent mechanical standards. Developments in power transmission, accuracy, and control of the movements and functions of the machine are constantly being incorporated into the design of new machine tools.

Machine tools turn metal into a variety of shapes, including cylindrical and conical surfaces, holes, plane surfaces, irregular contours, gear teeth, etc., as shown in Figure 1-1. Many machines, however, can produce a variety of surfaces. Thus, machine tools are built as general purpose machines, high production machines, and as special purpose machines. As the name implies, general purpose machine tools are designed to be quickly and easily adapted to a large variety of operations on many different kinds of parts, such as the type of projects a home hobbyist might tackle. Production machine tools are designed to perform an operation, or a sequence of operations, in a repetitive manner in order to achieve a rapid output of machined parts at minimum cost. Special purpose machine tools are designed to perform one operation, or a sequence of operations, repetitively, on a specific part. These machines are usually automatic and are unattended except when it is necessary to change and to adjust the cutting tools. They are used in mass-production shops such as are found in the automotive industry. CNC, or computer numerical controlled machines, are computer-aided machines. They are mainly used in high-production processes, but small machines used by the home hobbyist can also be CNC machines. See Figure 1-2.

Figure 1-1
An example of a machine tool creating a cylindrical shape from a metal bar.

Figure 1-2
A CNC benchtop mill.
Photo courtesy of Sherline.

Metal Cutting Tools

The heart of a shop machine is its metal cutting component—the metal cutting tool. Metal cutting tools separate chips from the workpiece in order to cut the part to the desired shape and size. There are a variety of metal cutting tools, each of which is designed to perform a particular job or a group of metal cutting operations in an efficient manner. For example, a twist drill is designed to drill a hole having a particular size, while a turning tool might be used to turn a variety of cylindrical shapes. In order to sever a chip from the workpiece the following conditions must be present:

1. The tool is harder than the metal to be cut.
2. The tool is shaped so that its cutting edge can penetrate the work.
3. The tool is strong and rigid enough to resist the cutting forces.
4. There must be movement of either the work or the cutting tool to make the cutting action possible.

Modern metal cutting tools are made from tool steels, powdered metals, ceramics, and industrial diamonds. These materials can be made to be very hard, and they can retain their hardness at the high temperatures resulting from the metal cutting action. All metal cutting tools wear as the result of stresses and temperatures encountered in separating the chips. The rate-of-wear must be controlled by the application of the correct cutting speed and feed. After wear has progressed to certain limits, the cutting edge may be resharpened by grinding. Ultimately, further sharpening is not practical and the tool must be discarded.

Figure 1-3 A drill boring into a metal surface.

*Figure 1-4
An abrasive wheel used for
tool sharpening.*

There are three basic types of metal cutting tools: single-point tools, multiple-point tools, and abrasives. These names are quite descriptive. A single-point metal cutting tool has a single cutting edge and is used for turning, boring, and shaping. Multiple-point tools have two or more cutting edges such as drills, reamers, and milling cutters. See Figure 1-3. Grinding wheels are an example of abrasive cutting tools, shown in Figure 1-4. Each grinding wheel has thousands of embedded abrasive particles which are capable of penetrating the workpiece and removing a tiny chip. The combined total of the tiny grinding chips can result in a substantial amount of metal being removed from the workpiece.

Planning the Home Workshop

There are many elements that determine the makeup of a home machine shop. They include the available space, budget, and the types of projects that will be completed in the shop. Because machines are the heart of any machine shop, *Machining for the Hobbyist* will examine the principal pieces of equipment that the home hobbyist may use—brief descriptions appear below with more detailed information about each piece appearing later in the book. Use this information as starting points to design and assemble your home shop.

Lathes

Lathes are machines that turn a workpiece while a stationary cutting tool removes metal from the work. See Figure 1-5. The action of the lathe allows the hobbyist to shape metal into

Figure 1-5 Benchtop lathes of different sizes.
Photo courtesy of Sherline

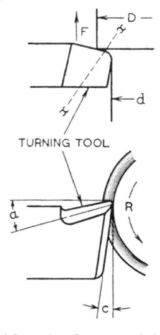

Fig. 1-6 Principles of turning and planing.

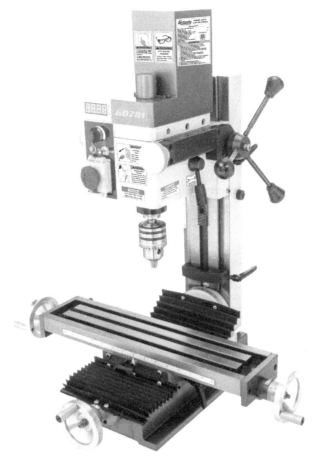

Figure 1-7
A vertical benchtop mill.
Photo courtesy of
Grizzly Industrial, Inc.

cylindrical shapes, bore holes and cut internal threads among other functions.

The diagram Figure 1-6 illustrates how metal is removed when turned on a lathe. The turning, or cutting, tool is held rigidly in some form of holder. The part to be turned rotates in direction R, and the tool feeds along continuously in direction F, parallel to the axis of the part being turned. Consequently, the part is reduced from diameter D, to some smaller diameter d, depending upon the requirements in each case. The upper surface of the tool, against which the chip bears as it is being severed, usually is ground to some rake angle a, to make the edge keener and to lessen cutting resistance. This top or chip-bearing surface generally slopes away from that part or section of the cutting edge which normally does the cutting; there is both backward and side slope or, in tool parlance, a back rake angle and a side rake angle. (The angle, a, of the diagram is intended to represent the combined back and side angles or what is known as the compound angle in some plane x-x.) If this inclination a, is excessive, the tool point will be weakened. On the other hand, if there is no inclination, the tool will still remove metal, but not as effectively as a tool with rake, especially in cutting iron and steel. Below the cutting edge, some relief angle c, is essential so that this end or side surface of the tool will not rub against the work and prevent the cutting edge from entering it freely, particularly when the tool is adjusted inward for taking a deeper cut. This relief is not confined to the end alone but is also required along the side extending as far back as the length of the cutting edge.

Mills

Unlike lathes, the cutting tools on milling machines rotate to remove metal from the workpiece. See Figure 1-7. The milling cutter represented by the upper left diagram, Fig. 1-8, has equally spaced teeth around its circumference so that as many cuts are taken per cutter revolution as there are teeth, and the cutting action is continuous. The cutter is rotated by the spindle of the milling machine and the part to be milled is given a feeding movement, generally in direction *F*, or *against* the cutter rotation *R*. (Sometimes the feeding movement is in the opposite direction, or *with* rotation R.) As the work feeds past the revolving cutter, the surface *H*, is reduced to some height *h*, as may be required.

The milled surface will *be* flat if the cutter is cylindrical, but, frequently, cutters of other forms are used. The chip-bearing surfaces of the milling cutter teeth may have some rake angle *a*, to lessen cutting resistance and they might have relief *c*, to permit free cutting action. Each tooth depends for its cutting action upon a relief angle and usually a rake angle also, just as with the turning and planing tools. (The actual relief angle *c*, applies to a narrow top edge on each tooth as indicated by the enlarged view, at upper right, of one tooth. Normally, the cutter is sharpened by grinding this narrow edge or "land" as it is called.)

The lower diagram in Figure 1-8 illustrates the action of a face milling cutter. This general type of cutter is designed especially for milling flat surfaces. To simplify the diagram, a cutter is shown with two cutting blades only. In actual practice, however, there would be a number of blades equally spaced around the circumference of the cutter body to obtain more continuous cutting action. As the cutter rotates in direction *R,* the work feeds past it as indicated by arrow *F,* so that each blade, as it sweeps around, reduces the surface from some height *H,* to *h*. The feeding movement of the work is not always in a straight line hut may be rotary. This, however, does not affect the operating principle. Each blade is held in a cutter body at an angle as indicated by the detail

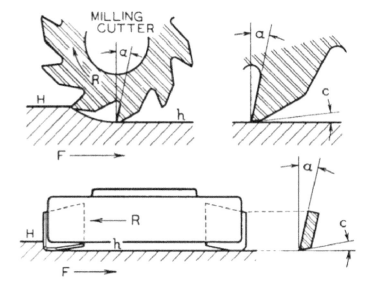

Figure 1-8 Principles of plain milling and face milling.

view to the right. This inclined position gives each blade a certain rake angle, a. The lower end of each blade also has a relief angle c. The face milling cutter is another illustration of the fact that metal cutting operations with different types and designs of tools are all based upon the same fundamental principles.

Drill Presses

Drill presses open and enlarge holes. They also finish a hole by reaming, boring, counterboring, countersinking and tapping. See Figure 1-9.

When a hole is drilled by using a twist drill, the cutting point, or end, is ground to provide a certain amount of relief c, in back of each cutting edge (left, Figure 1-10) and extending along the entire length of the two edges. The twisting, or helical-shaped flute, or groove through which the chips pass also provides rake angle a above each cutting edge. This rake, or backward slope, is away from the cutting edge and is also applied to the chip-bearing surface as in the case of the other tools. It will be understood that the drilling of a hole is

Figure 1-9 A drill press in operation.

done either by feeding a rotating drill in direction F, and into stationary work as when using a drilling machine, or by rotating the work itself instead of the drill as, for example, when using a lathe or turret lathe for drilling.

The diagram, Figure 1-10, right, illustrates a counterbore. This is a tool for enlarging part of a hole from some diameter d, to D. A pilot, or extension, fits into the smaller hole and ensures cutting the enlarged part concentric with it. Again we have on the counterbore, relief angle c, at the lower end of each cutting tooth, and also rake angles a. While the counterbore and the twist drill are in appearance unlike a milling cutter or turning tool, all of these tools conform to the same fundamental operating principles.

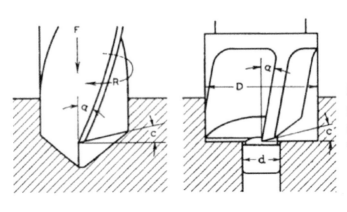

Figure 1-10
(Left) Cutting end of a twist drill.
(Right) Counterbore for enlarging a previously drilled hole.

Grinding Wheels

Grinding wheels are just as much metal cutting tools as are single- or multiple-point cutting tools which are used to alter the shape of a workpiece. In the industrial machine shop, grinding wheels are usually mounted on the spindles of precision machine tools in order to machine parts to close tolerances and to produce fine surface finishes. In the home shop, grinding wheels are used to sharpen cutting tools, among other functions.

This capability is further enhanced by the nature of the cutting action of the grinding wheel which produces a multitude of very small chips. These small chips permit very small amounts of metal to be removed from the surface of the work, thus favoring both dimensional accuracy and surface finish. Another important characteristic of the grinding wheel is that it will readily penetrate hardened metals, such as hardened tool steel, enabling these materials to be machined economically on precision grinding machines. See Figure 1-11.

Figure 1-11 A benchtop grinding wheel. Note the safety shields in front of each wheel.

Other Tools

In addition to the centerpiece tools mentioned above, there are literally hundreds of other tools available to the home hobbyist. Because machining requires a high degree of accuracy, many of the tools are used for measuring, laying out, and evaluating workpieces. These include calipers, micrometers, layout gages, and the like. See Chapter 2 for a look at some of the layout and measuring tools that will make your work more productive and efficient. Chapter 3 covers some of the more general tools that will come in handy for the home hobbyist. If you are just starting out and on a budget, purchase the tools you need for the projects you are attempting now. Buy quality tools because they make your work easier and safer.

Shop Layouts

Anyone reading this will have, or is planning to have, a home shop that is truly unique. The design and layout of each shop will depend on the available space, the budget, the tools and machines that will be part of the shop, and the projects that will be attempted there. The capabilities of the person who can devote an entire two-car garage to a shop will be different from the home machinist who has a corner of the basement to work with. Consequently, there are no set of general guidelines that apply to everyone. But there are some general rules-of-thumb that the novice will find helpful.

Shop Layout. If you are starting from scratch, draw a shop layout first. Draw the shape of the space to scale on graph paper. Then cut out pieces of color paper that represent the larger items you will include, such as a table for a tabletop lathe or a large cabinet for tools. You can also draw the footprint of the space on graph paper and then use sheets of tracing paper to experiment with different layouts. Some of the items to include are work areas and clearances around work areas, material and tool storage, and aisles for getting from one area to another. There are also computer applications for designing living spaces that can be adapted for shop design.

Lighting. Plan on two types of lighting: general lighting and specific task lighting. Overhead fixtures work well for providing general illumination. Light work areas with task lighting, which are usually wall- or ceiling-mounted spotlights you can aim, work lights, lamps, or any fixture that lets you train the light on the work.

Electrical. Plan on electrical service that is sufficient to power the equipment in the shop. This could take some figuring and may require the assistance of a licensed electrician. But, generally, it is best to have separate circuits for lighting and equipment. Provide electrical receptacles (outlets) around all work areas. Plan their locations so that you do not need to rely on extension cords. Be sure to follow the electric codes. In some areas that means having some or all the work performed by a licensed electrician.

Cleanup. Lathes and mills produce chips. It may make sense to place them near one another to make cleanup easier.

Storage. Even small shops need adequate storage, including places to put raw materials, hand tools, and parts.

Comfort. Work benches should be a comfortable height for you. If you are not sure what would be comfortable, consider that a standard kitchen counter is 36 inches from the floor. And don't forget to include stools in your plan. Heating and cooling are other component to consider. An un-heated garage shop in the north won't get much use during the winter.

Room to Grow. Most home shops are never big enough. Even if you start out with plenty of elbow room, additional machines and tools will make the area shrink. If possible, try to build room to expand in your plans.

Shop Safety

While home shops do not present as many hazards as industrial machine shops, they do contain tools and equipment that can cut metal. It does not take much to imagine what they can do to skin and bone. Fortunately, most machine shop safety is really a matter of common sense. Here are a few things to keep in mind.

Protect Your Eyes. Wear safety glasses while working—not only when cutting and filing metal, but also when using solvents.

Dress Appropriately. Avoid loose fitting clothes, as well as watches and other jewelry. While gloves may be appropriate for some tasks, don't wear them when they can get caught in moving parts of machines. Tie back long hair or keep it under a cap.

Read the Manual. Follow the manufacturer's instruction regarding setup, operation, maintenance, and safety of all equipment. Keep tools and equipment in good repair.

Pay Attention. Avoid all distractions while working, and keep your mind on the job. Don't work and talk; don't work and text; don't work and watch TV.

Lock It Up. Keep the shop locked when you are not there.

Must Haves. The shop should include a first-aid kit and a fire extinguisher.

2   Measuring Tools

Measuring tools are used in machine shops in order to secure the required sizes and degree of accuracy of the parts. There are many different types and designs of measuring tools, some of which are very sophisticated. This chapter will deal with the measuring tools normally used by machinists and home hobbyists.

All precision measuring tools must be handled with care in order to avoid the slightest damage to them. Measurements accurate to one thousandth (.001) inch, or even to one ten-thousandth (.0001) inch, must often be made. Careless handling of precision measuring tools will destroy their accuracy, and hence their usefulness, in making exact measurements. For example, a square that is not a true square will result in spoiled work. Precision measuring tools, therefore, must frequently be checked in order to verify their accuracy. With many of these tools the machinist must depend upon the sense of feel in order to make accurate measurements. This requires that they be correctly grasped by the hands in order to utilize the sensitive nerves that are located in the tips of the fingers. The proper manipulation of precision tools is shown in many of the illustrations in this chapter. See Figure 2-1.

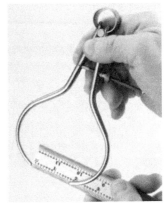

Figure 2-1 In some cases, more than one tool is needed. Here a caliper is used with a steel rule to take a measurement.
Courtesy of The L.S. Starrett Company

The Steel Rule

The rule is a basic length-measuring tool. It may be used directly in measuring a length; or, used indirectly as when the diameter of a cylindrical object is measured with outside calipers and this measurement is then transferred from the calipers to the rule. Precision-made, hardened steel rules work best in the machine shop. The lines or graduations on these rules are machine divided on specially designed, and very precise, graduating machines. Precision steel rules may be obtained as spring temper, semispring temper, semi-flexible, or full flexible. They range in size from one-quarter inch in length for measuring in grooves, recesses, and key-seats to twelve feet in length for large work. The most frequently used lengths are the six-inch and twelve-inch rules.

Two systems of graduations (Figure 2-2) are used on steel rules. They are the decimal inch system and the fractional inch system. In the decimal inch system, the inch is divided into one tenth (.100) inch, one-fiftieth (.020) inch, and one-hundredth (.010) inch. (See upper view, Figure 2-2.) In the fractional inch system (lower view, Figure 2-2), the inch is successively divided by two, yielding the following graduations: ½ or .500 inch, ¼ or .250 inch, 1/8 or .125 inch, or 1/16 .0625 inch, 1/32 or .03125 inch, and 1/64 or .015625 inch. One advantage of the fractional inch system is that the smallest graduation that can be clearly distinguished with normal eyesight; without causing undue eyestrain, it is 1/64 or .0156 inch as compared to 1/50 or .020 inch for the decimal inch system.

Decimal equivalents are the decimals corresponding to the fractions in the fractional inch system. The decimal equivalent can be calculated from the fraction by dividing the denominator into the numerator. For example, to find the decimal equivalent of 7/32 inch, divide 7 by 32, thus $7 \div 32 = .21875$ inch. Tables of decimal equivalents are usually available, making these calculations unnecessary. Their use is recommended to prevent possible mistakes that sometimes are made when decimal equivalents are found by calculation. A table of decimal equivalents is given in the Appendix of this text.

Figure 2-2 (Upper view) Steel rule with decimal inch graduations;
(Lower view) Steel rule with fractional inch graduations.

Courtesy of The L.S. Starrett Company

Metric Steel Rules

The graduations on metric rules are in millimeters on one scale and one-half millimeters on the other. Usually each ten millimeter length on the rule is numbered on consecutive sequence. To compare the smallest graduation on metric and inch rules: 1.0 mm is equal to .039 in. Metric rules are made in lengths of 150 mm (approximately equivalent to a 6-inch rule), 300 mm, 500 mm, and 1000 mm. Steel rules are available having metric graduations on one side and inch graduations on the other side. See Figure 2-3.

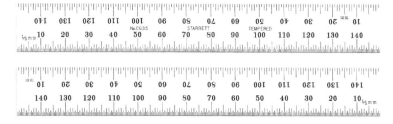

Figure 2-3 Steel rule with metric graduations.

Telescoping Gage and Small-Hole Gages

The telescoping gage has two contact plungers which expand outwardly across the diameter of a hole when the knurled nut at the end of the gage stem is released. When both plungers have contacted the hole, this nut is again tightened and the telescoping gage measurement is transferred to outside micrometers. Small-hole gages have two contact points that are expanded by turning the knurled nut on the end of the stem until the gage can pass through the hole with a very light contact "feel." The hole measurement is then transferred to outside micrometers. See Figure 2-4.

Figure 2-4 Small-hole gages are used to measure hole diameters.

Courtesy of The L.S. Starrett Company

Calipers

There are three basic kinds of calipers: 1. Outside calipers, used primarily to measure outside diameters; 2. Inside calipers, used primarily to measure hole diameters; and 3. Hermaphrodite calipers, used to measure the distance between a surface and a scribed line or to scribe a line from a surface. Two types of construction, firm joint and spring, are used in calipers. Firm joint calipers utilize the friction between the legs to maintain their setting. A firm joint hermaphrodite caliper is shown in Figure 2-5. Spring calipers have a curved spring at the upper, or pivot, end which forces the caliper legs against the adjusting screw nut, thereby maintaining the caliper adjustment.

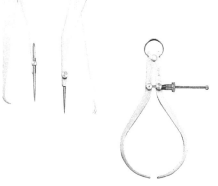

Figure 2-5
The three types of calipers from left:
inside calipers, hermaphrodite calipers,
and outside calipers.
Courtesy of The L.S. Starrett Company

Using Calipers

Accurate measurements are obtained with outside calipers by holding them as shown at A, in Figure 2-6, and passing them over the workpiece with a very light contact pressure that can just be felt by the fingertips. The caliper reading is then transferred to a rule (Figure 2-6, B). Accurate caliper readings, however, cannot be made when the work is rotating. In measuring a hole, inside calipers are adjusted to the correct diameter of the hole (see A, Figure 2-7) by holding one caliper leg against the side of the hole with a finger of one hand, while the caliper, held in the other hand, is carefully passed through the hole. Moving the free leg back and forth in the X direction (A, Figure 2-7), while slowly passing through the hole in the Y direction, will assist in finding the smallest true diameter of the hole through which the caliper legs can pass. Again, a very light feel will determine this setting. The inside caliper measurement is then transferred to a steel rule (B, Figure 2-7) or to an outside micrometer (D, Figure 2-7). When done with great care, measurements accurate to "tenths" (.0001 inch) can be made, using both inside calipers and outside calipers. Measurements can also be transferred between inside and outside calipers as shown at C, in Figure 2-7.

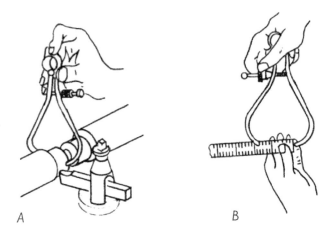

A   B

Figure 2-6 Correct method of using outside calipers.

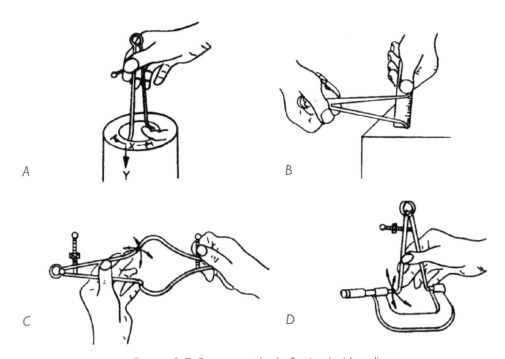

A   B

C   D

Figure 2-7 Correct method of using inside calipers.

The Square and the Bevel Protractor

A frequently encountered angle in machine shop work is the right angle, or 90 degree angle. The square is the basic tool used to measure and test for this angle. For very accurate determinations of the right angle, a master precision square, Figure 2-8, is used. The beam and the blade of this square are hardened, ground, and lapped to insure straightness, parallelism, and the exact, right-angle relationship. It should be very carefully handled at all times.

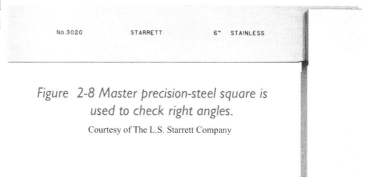

Figure 2-8 Master precision-steel square is used to check right angles.

Courtesy of The L.S. Starrett Company

A combination set is shown in Figure 2-9. Attached to a precision steel rule, from left to right, are a square head, a bevel protractor, and a center head. These heads can be changed to any desired position along the length of the rule, or they may be removed completely. The square head and the bevel protractor both contain spirit levels which are often of assistance in making measurements. The square head has a short edge machined at 45 degrees with respect to the right angle. The center head provides a way to find the center of both cylindrical and square work.

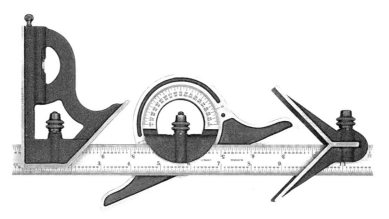

Figure 2-9 Combination set including square head, bevel protractor, and center head.

Courtesy of The L.S. Starrett Company

Checking Squares for Square

Squares may be checked by using two toolmakers buttons that are screwed to an angle plate as shown in Figure 2-10. The lower button is screwed tight, while the other button is tightened also, but less firmly. The square and the angle plate are then set on a precision-surface plate and finally, the blade of the square is placed against the buttons. A thin piece of paper, or paper "feeler," is placed between the blade and each button. The upper button is then lightly tapped with a piece of soft metal, say a 1-inch dia. × 6-inch piece of soft brass, until both paper feelers are tight. The square is then positioned on the other side of the buttons and checked with paper feelers as before. If both paper feelers are tight, the square is accurate. In making this test, it is very important that both toolmakers buttons have exactly the same diameter. The thin paper feelers are often much more sensitive than eyesight when making precise angular measurements.

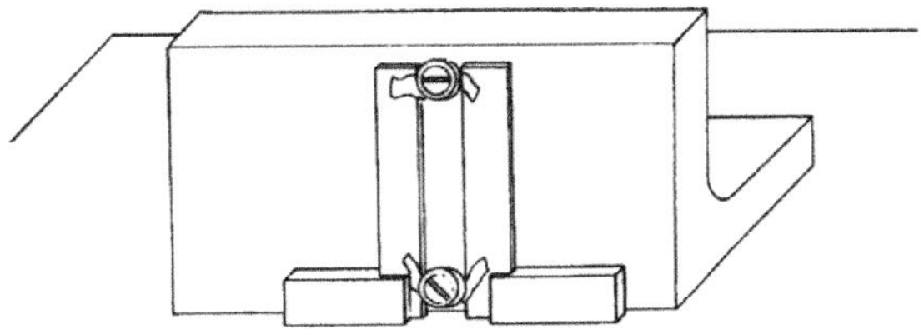

Figure 2-10
Method for checking a square.

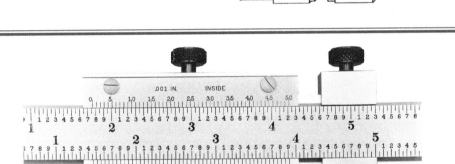

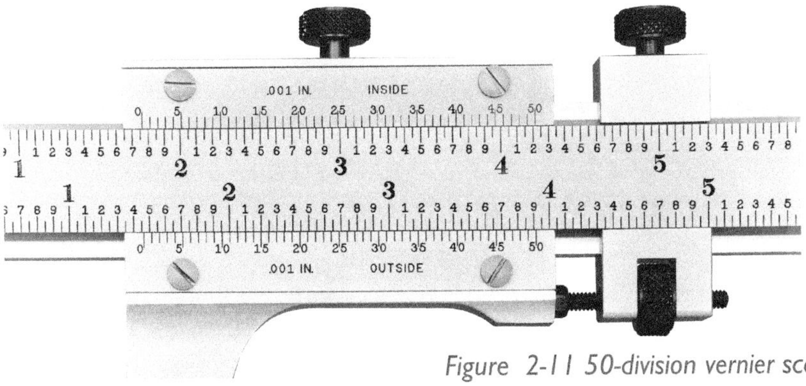

Figure 2-11 50-division vernier scale.

Vernier Measuring Instruments

The vernier is an auxiliary scale that is attached to calipers, height gages, depth gages, protractors, and other measuring instruments. It allows exceedingly accurate measurements to be made. The vernier measuring instrument is, in effect, a graduated steel rule to which a sliding member containing the vernier scale is attached. There are two types of vernier scales commonly used in the machine shop; one has a 25-division vernier scale and the other has a 50-division vernier scale.

Reading a Vernier Scale

To read a 50-division vernier follow the steps below and refer to Figure 2-11:

Read the whole inches on the primary scale	1.000
Read the nearest small number to the left of the 0 indicator line and multiply times .100, or 4 × .100	.400
Count the number of small graduations beyond the previous reading and multiply times .050, or 1 ×.050	.050
Find the graduation on the vernier scale that is in exact alignment with a graduation on the primary scale and multiply times .001, or 14 × .001	.014
Add total to obtain reading	1.464 inches

Other Types of Calipers

Vernier slide calipers are capable of measuring both inside and outside diameters. The tool consists of steel rule, which has a fixed jaw at one end and a sliding graduated scale, which also has a jaw attached to it. (Figure 2-12). When the jaws are aligned on the workpiece, the scale is read to achieve an extremely accurate measurement. Newer more advanced tools feature a dial or a digital electronic readout. See Figures 2-13 and 2-14.

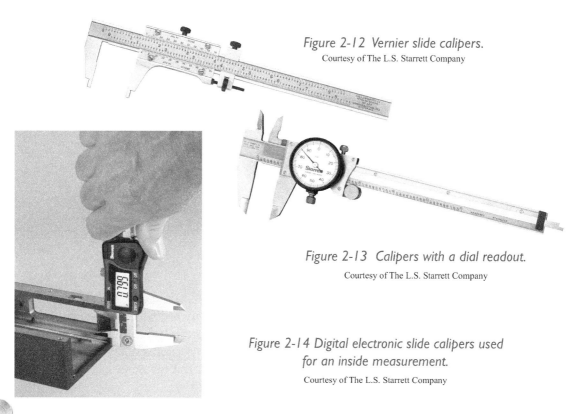

Figure 2-12 Vernier slide calipers.
Courtesy of The L.S. Starrett Company

Figure 2-13 Calipers with a dial readout.
Courtesy of The L.S. Starrett Company

Figure 2-14 Digital electronic slide calipers used for an inside measurement.
Courtesy of The L.S. Starrett Company

Vernier Height Gages

Vernier height gages (Figure 2-15A) are used to measure vertical distances above a reference plane; usually a precision-surface plate. They are also used to scribe lines at a given distance above a reference plane when making precise layouts on workpieces. A vernier depth-gage is shown at B in Figure 2-15.

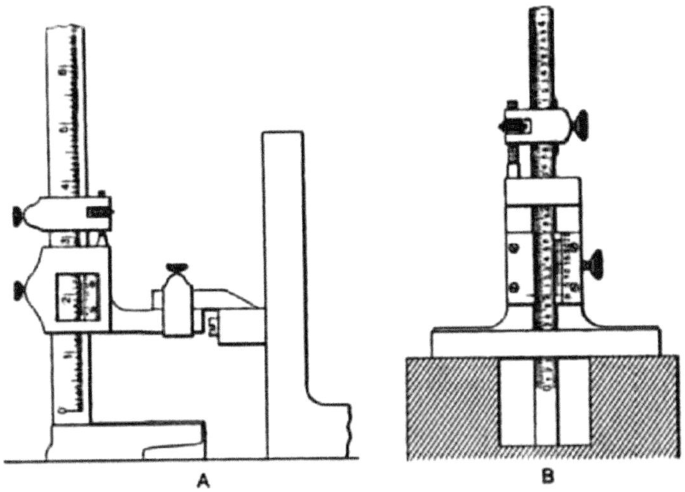

Figure 2-15
A. Vernier height-gage.
B. Vernier depth-gage.

A   B

Figure 2-16 A vernier height gage.

Courtesy of The L.S. Starrett Company

19

Bevel Protractors

Vernier bevel protractors, as shown in Figure 2-17, are used to measure angles to an accuracy of five minutes (05'). Since there are sixty minutes (60') in one degree, this is equal to one-twelfth (5/60 =1/12) of a degree. The main protractor scale is divided into degrees, with every ten degrees numbered. The vernier scale is actually two vernier scales, each having twelve divisions on either side of the zero graduation. The left-hand scale is used when the vernier zero graduation is moved to the left of the zero on the primary scale, while the right-hand scale is used when the movement is to the right. Figure 2-18 shows a vernier protractor used in combination with a height gage. This type of tools allows any angle to be laid out or measured with great accuracy.

Figure 2-17 Vernier bevel protractor.

Courtesy of The L.S. Starrett Company

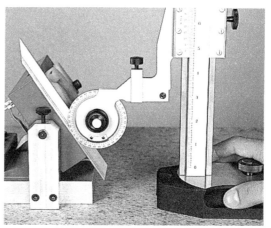

Figure 2-18 A vernier bevel protractor used in combination with a height gage.

Courtesy of The L.S. Starrett Company

Reading the Vernier Bevel Protractor

To read the vernier bevel protractor, first read the number of whole degrees passed by the vernier zero and then count, in the same direction, the number of graduations between the vernier zero and that line which exactly coincides with a graduation on the primary or degree scale; this number multiplied by 5 will give the number of minutes to be added to the whole number of degrees. In Figure 2-19, the vernier zero has passed to the left of 50°, and the fourth line to the left of the vernier zero coincides with a line on the degree scale, (see the stars in the illustration). Hence, the reading is 50° + 4 × 5' or 50°20'.

Figure 2-19 Vernier bevel protractor scale.

Courtesy of The L.S. Starrett Company

Metric Vernier Measuring Instruments

Metric vernier height gages, vernier calipers, and other vernier measuring instruments read to an accuracy of 0.02 mm, which is equal to.00079 inch. Like inch reading verniers, there are two types of metric vernier scales, 25-division and 50-division scales. Some vernier measuring instruments have two sets of graduations, one in inch units and the other in millimeters.

The metric vernier scale has 50 graduations, each representing 0.02 mm. Every fifth graduation is numbered in sequence 0.10 mm, 0.20 mm, 0.30 mm, etc. To read a 50-division metric vernier, follow the steps below and refer to Figure 2-20.

1. Read the numbered graduation on the primary metric scale to the left of the 0 on the vernier scale. For an outside reading, use the bottom scale. 20.00

2. Count the number of graduations between the numbered graduation on the primary scale and the zero graduation on the vernier scale. Each graduation is one millimeter. 7.00

3. Find the graduation on the vernier scale that exactly coincides with a graduation on the primary scale; then read the vernier scale. Each vernier graduation represents 0.02 mm. In the illustration the vernier scale reading is 0.42 mm. 42

4. Add together to obtain the answer. 27.42 mm

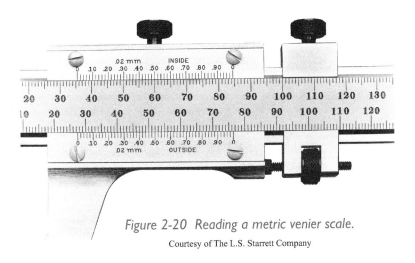

Figure 2-20 Reading a metric venier scale.

Courtesy of The L.S. Starrett Company

Micrometer Measuring Instruments

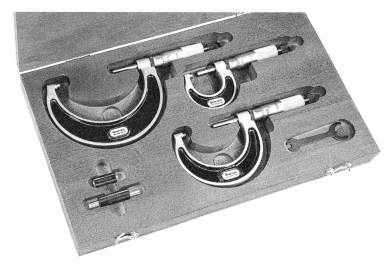

Figure 2-21A
A set of outside micrometers.

Courtesy of The L.S. Starrett Company

There are micrometers to take inside, outside, and depth measurements. While the tools are designed differently, the measuring device is the same for all. The micrometer screw principle is used on a variety of measuring instruments which includes outside micrometers, inside micrometers, and micrometer depth gages. The precision obtainable by all micrometer measuring instruments is dependent upon the micrometer screw; therefore, the thread of this screw is made with the greatest possible care and degree of precision. For this reason, micrometer screw threads are very seldom made longer than one inch and micrometer measuring instruments are designed to accommodate a one-inch movement of the micrometer spindle. For example, outside micrometers are made with frames having sizes of one-inch increments. See Figure 2-21A.

Figure 2-21B illustrates a sectional view of a one-inch, outside micrometer with all of the parts named. The principal parts are the frame, sleeve, thimble, spindle, and anvil. The frame supports the components of the micrometer and keeps them in correct alignment.

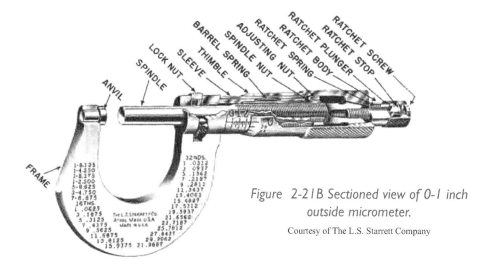

Figure 2-21B Sectioned view of 0-1 inch outside micrometer.

Courtesy of The L.S. Starrett Company

The sleeve houses the spindle nut and it has graduations marked lengthwise on its outside surface. See Figures 2-22 and 2-23. These graduations are one-inch long, and they are divided into .100-inch and .025-inch intervals. The thimble is attached to the spindle by a taper and a screw. The thimble rotates and moves lengthwise over the sleeve. One end of the thimble is beveled. The bevel on the thimble has 25 equally spaced graduations around its circumference. The spindle has the micrometer thread on one end and one of the two measuring surfaces on the face of the other. The other measuring surface is on the face of the anvil. These two measuring surfaces are made to be exactly parallel to each other regardless of the position of the spindle.

The micrometer screw thread has 40 threads per inch. Its lead, or distance that the thread advances per revolution, is equal to 1 ÷ 40 or .025 inch or the distance between the smallest graduations on the sleeve. The beveled edge of the thimble will move lengthwise one graduation on the sleeve for each revolution, since it is attached to the screw. Turning the thimble through one graduation on the thimble scale will cause the

Figure 2-22 *Micrometer with an electronic readout.*

Courtesy of The L.S. Starrett Company

thimble and spindle assembly to rotate exactly 1/25 of a revolution, since there are 25 equally spaced graduations around the thimble. When the spindle rotates 1/25 of a turn, it moves lengthwise 1/25 of its lead, or $1/25 \times .025 = .001$ inch.

Figure 2-23 *Taking a measurement with an inside micrometer.*

Courtesy of The L.S. Starrett Company

Reading a Micrometer

To read a micrometer scale follow the steps below and refer to A in Figure 2-24.

Read the largest number visible on the sleeve scale and multiply by .100, or 2 × .100	.200
Count the number of small graduations beyond the graduation corresponding to the number read above, and multiply by .025, or 3 × .025	.075
Read the graduation on the thimble scale and multiply by .001, or 10 × .001	.010
Add to obtain reading	.285 inch

Some micrometers have a vernier scale, v, marked on the sleeve as shown at B, Figure 2-24. The vernier scale, in conjunction with the regular micrometer scale, enables the micrometer to be read to .0001 inch. The relation between the graduations on the vernier and those of the regular micrometer is more clearly shown in C, in Figure 2-24. The vernier has ten divisions which are the same length as nine graduations on the thimble. For convenience in reading, each graduation on the vernier scale is numbered.

The difference in width of a thimble division and a vernier division is equal to one-tenth of the thimble division. Therefore, a movement of the thimble equal to this difference results in a movement equal to 1/10 of 1/25 of a revolution, or 1/10 × 1/25 = 1/250 of a turn of the thimble and spindle assembly. This will cause the spindle to move lengthwise 1/250×.025, or .0001 inch.

In order to read a micrometer having a ten-thousandths vernier scale, first determine the reading in thousandths, as with an ordinary micrometer, and then find a line on the vernier scale that exactly coincides with one on the thimble; the number on this line represents the number of ten thousandths of an inch to be added to the number of thousandths obtained by the regular micrometer scales. For example, at C, in Fig. 2-24, the micrometer scales read 0.275 inch. On the vernier scale, the fourth graduation coincides with a graduation on the thimble; therefore, 4 ×.0001, or .0004 inch is added to the regular micrometer reading. The final reading is, then, .275 +.0004, or .2754 inch.

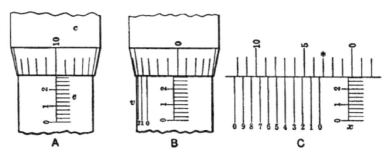

Figure 2-24 Micrometer graduations.

Metric Micrometers

Metric micrometers are available with or without vernier scales. Those without have an accuracy of 0.01 mm (.00039 inch). Those with a vernier scale will read in 0.002 mm (.00008 inch increments. The usual range of each metric micrometer is 0 to 25 mm, corresponding to 0 to .984 inch; metric micrometers are available in sizes up to 600 mm and larger, if required.

The reading line on the sleeve is graduated in one millimeter (1.00) increments above the line, and below the line each millimeter is divided in half by a graduation. Therefore, the reading line is divided into one-half millimeter increments by the graduations above and below the reading line. Every fifth line above the reading line is numbered from 0 to 25, indicating five millimeter (5.00) intervals between the numbered lines.

The pitch of the spindle screw on metric micrometers is one-half millimeter (0.50 mm). One revolution of the spindle will, therefore, advance the spindle 0.50 mm toward or away from the anvil. Since the thimble moves with the spindle, it will also move 0.50 mm, which is equal to the distance between graduations on the reading line.

The beveled edge on the thimble is graduated into 50 divisions, every fifth line being numbered. Each division represents a rotation of 1/50 of a revolution by the thimble and the spindle screw. Since one complete revolution of the thimble and spindle screw causes a movement of 0.50 mm by the spindle, 1/50 of a revolution will cause a movement of 1/50 × 0.50 mm, or 0.01 mm. Thus, one graduation on the thimble equals 0.01 mm, two graduations 0.02 mm, three graduations 0.03 mm, etc.

Reading a Metric Micrometer

To read the metric micrometer, add the number of millimeters and half-millimeters visible on the sleeve to the number of hundredths of a millimeter indicated by the thimble graduation coinciding with the reading line on the sleeve. To illustrate this procedure refer to the reading on the micrometer scale in Figure 2-25 and follow the steps below.

Read the large number visible on the reading line and express as whole millimeters. 5.00

Count the number of graduations visible above and below the reading line that are visible beyond the large number and multiply by 0.5, or 1 × 0.5   .50

Read the graduation on the thimble coinciding with the reading line and multiply by 0.01, or 28 × 0.01   .28

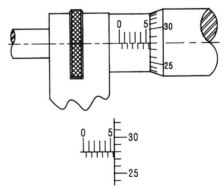

Add to obtain the reading. 5.78 mm

Figure 2-25 Reading a metric micrometer scale graduated to read to 0.01 mm.

Speciality Micrometers

The Starrett Mul-T-Anvil micrometer in Figure 2-26 can make a variety of measurements that cannot be made with a regular micrometer; e.g., it can measure the wall thickness of tubing, the length of a shoulder, and the distance from a hole or slot to an edge. Two anvils are furnished with this micrometer: one is cylindrical in shape and the other is in the form of a stepped flat having a thickness of approximately .125 inch on one end and .060 inch on the other. Other special anvils are available. The anvils can be quickly interchanged by simply loosening the clamping vise jaw.

By removing the clamping vise jaw and the clamping screw, the Mul-T Anvil micrometer can be converted into a micrometer height gage. Its range of measurement as a height gage can be extended beyond the one inch spindle travel of the micrometer by placing it on precision parallels or precision gage blocks having a known height.

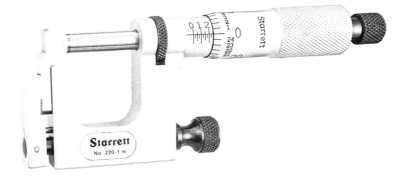

Figure 2-26
Mul-T-Anvil micrometer
can be used as
a micrometer or
as a height gage.

Courtesy of The L.S. Starrett Company

Checking Micrometers

All micrometers should occasionally be checked for accuracy and, if necessary, adjusted to read correctly. Zero-to-one-inch micrometers can be checked by noting if the micrometer reading is zero when the spindle touches the anvil. Both the spindle and the anvil should be cleaned beforehand by pulling a clean strip of paper between the anvil and the spindle. If a zero reading is not obtained, the sleeve should be adjusted by following the instructions supplied with the micrometer by the toolmaker. See Figure 2-27. Larger micrometers are checked in a similar way by placing a standard disc or pin between the anvil and the spindle and reading the micrometer. Standard disc and pin gages have a very precise known diameter or length, which is measured by the micrometer in making the check. They are usually furnished with the larger size micrometers, although they can also be obtained separately. Again, the contact surfaces of the micrometer and the gages must be clean before making this check.

A more precise check can be made on micrometers by measuring over precision gage blocks. (Precision gage blocks are described later on in this chapter.) The measurement obtained can be compared to the known size of the gage blocks. By measuring over a graduated series of different gage blocks, an indication of the accuracy of the micrometer screw can be obtained. Unless the micrometer is abused, the micrometer screw will wear to the extent of being imprecise on only extremely rare occasions. If this does happen, the micrometer should be replaced. On rare occasions, the adjusting nut on the micrometer screw must be adjusted. Micrometer depth gages are checked by measuring from the top of a stack of precision gage blocks to the surface of a precision surface plate. The measurement obtained is compared to the known length of the gage blocks.

Inside micrometers are usually checked with outside micrometers that are known to be precise. The inside micrometer is placed against the anvil. Then the spindle of the micrometer is carefully adjusted to read the length of the inside micrometer while it is moved about very slightly to obtain the feel of the correct setting. The readings of the two micrometers are compared to make the check. Sometimes inside micrometers are checked with precision gage blocks and, when available, with precision ring gages.

Figure 2-27
Adjusting for a zero reading with a spanner wrench supplied by the tool manufacturer.

Taking Accurate
Measurements

Accurate measurements made with outside micrometers depend upon the way in which they are handled. In order to feel when they are in correct contact with the workpiece, they must be held correctly when taking different types of measurements. Any movement of the frame, while manipulating the micrometer screw, should be kept to a

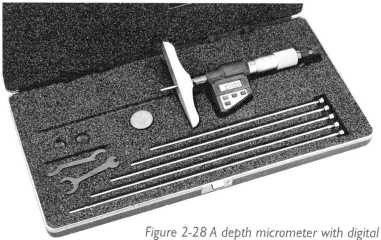

Figure 2-28 A depth micrometer with digital readout and assorted measuring rods.

Courtesy of The L.S. Starrett Company

minimum —just enough to establish the right feel. Inside micrometers, must be manipulated through a hole, or recess, much in the same manner as inside calipers, in order to find the true diameter of the hole. It is advisable to check the accuracy of inside micrometers with outside micrometer before they are used. The thimble on a depth micrometer is carefully turned until the measuring rod just touches the surface to be measured. See Figure 2-28.

Dial Test Indicators

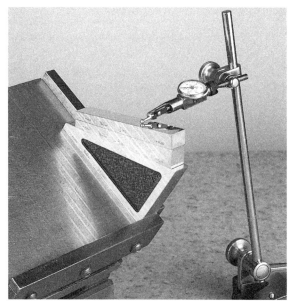

Figure 2-29   Dial test indicators serve a variety
of uses in the machine shop.
Here an indicator is used with a suface gage.

Courtesy of The L.S. Starrett Company

The dial test indicator is a sensitive instrument having a graduated dial (Figure 2-29) and an indicating hand which is connected through a system of multiplying levers, or gears, to a contact point that is fastened to the end of a spindle. The contact point is placed in contact with the workpiece. Any movement of the contact point is transmitted through the spindle, greatly amplified by the internal works, and displayed by a movement of the indicating hand on the dial. Dial test indicators are designed to read to .001 inch, .0005 inch, or .0001 inch. The reading of the divisions is usually marked on the dial face. They are made in a great variety of sizes and shapes. Indicators are widely used in machine shops in combination with many other forms of gaging and measuring devices. In the shop, dial test indicators are frequently simply referred to as "indicators."

Dial test indicators check the depth of cut on a flat casting. Among other applications of indicators are these: to locate edges and holes for precision boring, to measure concentricity, to "zero" the position of machine tool slides, and to measure parallelism between surfaces. The operation of using a dial test indicator is usually called "indicating."

Precision Gage Blocks

Precision gage blocks, Figure 2-30, are small blocks of steel that have been heat treated to a high hardness. The heat treatment is also designed to obtain a high degree of dimensional stability in the blocks. They are made to have an extremely high degree of dimensional accuracy as shown in Table 2-1. There are several grades of gage blocks which can be categorized as master blocks, inspection blocks, and working blocks. Working blocks are used in the workshop. They are used to calibrate micrometers and other precision measuring tools, to inspect dies, tools, and other precision machine parts. They are used with a height gage and the dial test indicator to measure the height of a surface on the workpiece by comparing the height of the gage block and the workpiece with the indicator. Gage blocks are also used to measure angles. Angles are sometimes machined on sine plates which are set by using gage blocks. Precision, angular gage blocks are also made for making angular measurements.

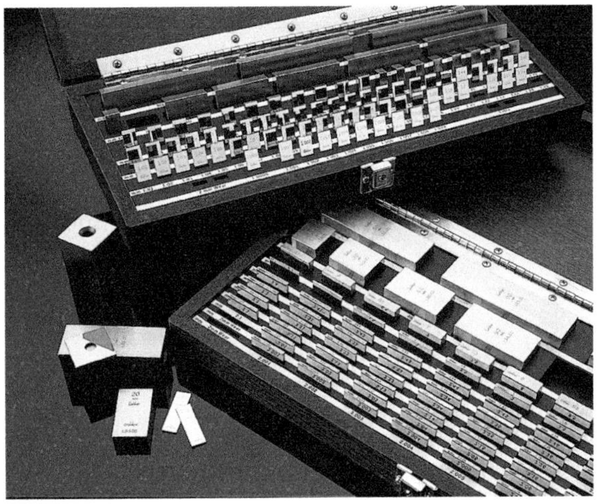

Figure 2-30 Among other uses, percision gage blocks help calibrate machine shop tools.

Courtesy of The L.S. Starrett Company

Gage blocks are sold in sets. A set of gage blocks is shown in Figure 2-30. The sizes for an 83-piece, gage-block set are given in Table 2-2. By combining two or more gage blocks, a large range of extremely accurate dimensions can be obtained. The blocks can be combined in the following way when there is in the set no single block of the exact size that is wanted. Two or more blocks are assembled, or combined, by wringing the blocks together to form the equivalent of a single block of a given dimension. First, the joining surfaces of the blocks are cleaned. One block is then placed over the other at an angle and the two are twisted, or wrung, together. When properly wrung together, the blocks will adhere to each other so strongly that a very considerable force is required to separate them. They are best separated by twisting them apart in a manner similar to the way in which they were assembled.

Table 2-1. Accuracy Standards for Precision Gage Blocks

Classification	English Measure Standards			Surface Finish	Metric Measure Standards		
	Length	Flatness	Parallelism		Length	Flatness	Parallelism
	inches			microinches	millimeters		
U.S. Federal AA	± .000002	.000004	.000003	.5	± .00005	. . .	. . .
British Reference Grade	± .000002	.000003	.000003		± .00005	.00008	.00008
U.S. Federal A	+ .000006 − .000002	.000004	.000004	.8	+ .00015 − .00005	. . .	. . .
British Calibration Grade	± .000005	.000003	.000003	. . .	± .00012	.00008	.00008
German Grade O	. . .	. . .	. . .	...	± .00014	.00010	.00010
U.S. Federal B	+ .000010 − .000006	.000006	.000005	1.2	+ .00025 .00015	. . .	. . .
British Inspection Grade	+ .000007 − .000003	.000005	.000005	. . .	+ .00020 − .00010	.00010	.00010
German Grade I	. . .	. . .	. . .	. . .	± .00030	.00015	.00015
U.S. Federal	. . .	. . .	. . .	. . .	. . .	. . .	. . .
British Workshop Grade	+ .000010 − .000005	.000010	.000010	. . .	+ .00025 − .00010	.00025	.00025
German Grade II	. . .	. . .	. . .	. . .	± .00070	.00025	.00025
German Grade III	. . .	. . .	. . .	. . .	± .00140	.00050	.00050

U.S. Federal Specification GGG-G-15.—British Standards 888-1950.—German Standards DIN 861.

Using Gage Blocks

In selecting gage blocks for a given dimension, the least number of blocks that will give the required dimension should be used. This is accomplished by successively eliminating the smallest remaining dimension.

Example 2-1.

Determine the gage blocks required to obtain 3.6742 inches (Refer to Table 2-2.)

Table 2-2. Sizes for an 83-Piece Gage-Block Set

First: .0001 Series (9 Blocks)								
.1001	.1002	.1003	.1004	.1005	.1006	.1007	.1008	.1009
Second: .001 Series (49 Blocks)												
.101	.102	.103	.104	.105	.106	.107	.108	.109	.101	.111	.112	.113
.114	.115	.116	.117	.118	.119	.120	.121	.122	.114	.124	.125	.126
.127	.128	.129	.130	.131	.132	.133	.134	.135	.127	.137	.138	.139
.140	.141	.142	.143	.144	.145	.146	.147	.148	.149			
Third: .050 Series (19 Blocks)											
.050	.100	.150	.200	.250	.300	.350	.400	.450	.500	.550	.600
.650	.700	.750	.800	.850	.900	.950					
Fourth: 1.000 Series (4 Blocks)				
		.000	2.000	3.000	4.000
Two .050 Wear Blocks					

Step 1. Eliminate the .0002 inch by selecting a .1002 block. Subtract .1002 from 3.6742. 3.6742 – .1002   3.5740

Step 2. Eliminate the .004 inch by selecting a .124 block. Subtract .124 from the remainder in Step 1 3.574 – .124   3.450

Step 3. Eliminate the .450 inch with a .450 block. Subtract .450 from the remainder in Step 2. This obviously leaves 3.000, which is obtained with a 3-inch block.

Step 4. Check the blocks to determine if they will result in the required dimension .1002 + .124 + .450 + 3.000   3.6743 inches

Surface Gages

The surface gage, Figure 2-31, is one of the most useful tools employed by the machinist. On certain classes of work, it is almost an indispensable tool. For example, surface gages can be used to scribe layout lines on workpieces in order to locate the centers of holes, to provide reference lines for setting up workpieces on the machine, or to outline surfaces to be machined. In Figure 2-31, the scriber point of the surface gage is being positioned by a rule in preparation for scribing a line on the workpiece. The rule is conveniently held in a vertical position by a rule holder; a surface plate provides a reference plane for transferring the measurement from the rule to the workpiece by means of the surface gage. Located at the end of a rocker plate, a vertical thumb screw provides a very fine adjustment to the spindle and allows accurate settings of the scriber to be made.

A dial test indicator, which can be used to transfer precise measurements, can be attached to the spindle or to the scriber. The scriber point can be used to check the position of layout lines when setting up a workpiece on the machine. The bent point of the scriber can be used to lay out lines that are close to the base of the surface gage. It is also used to test the parallelism between a surface on the workpiece and the top of a surface plate or machine tool table. This can be done whether the surface of the workpiece has been finished smooth or if it is still rough. The use of the surface gage to test a cast surface for parallelism is especially effective because the surface gage can average out the irregularities on such surfaces.

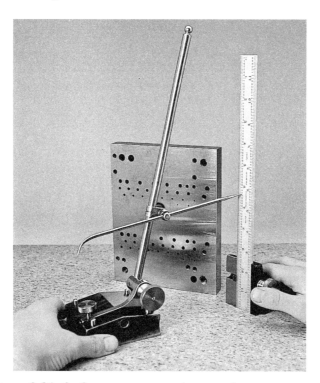

Figure 2-31 Surface gages are used to transfer measurements.
Courtesy of The L.S. Starrett Company

Parallels

While not primarily a measuring tool, parallels are indispensable tools in machine shop and tool room work. They are used for the setup of work pieces on machine tools, when inspecting parts, and when making layouts on parts. Parallels are rectangular bars of hardened or unhardened steel that have two sides ground exactly parallel; some parallels have two pairs of sides ground parallel. Frequently parallels are made in pairs, each pair being ground together so that they are exactly the same size. Very large parallels are usually made of alloy cast iron; highly precise parallels are also made from granite because this material is very stable dimensionally. Hardened steel precision parallels have two pairs of sides parallel, the sides being square and parallel to "tenth" (.0001 inch or 0.002 mm) tolerances. These parallels are made in fractional inch sizes ranging from 1/4 × 3/8 ×6 inches to 1 1/2 × 3 × 12 inches; metric sizes range from 6 × 10 × 150 mm to 38 × 76 × 300 mm. Special sizes are often used, when needed. Although not classified as parallels, 1-2-3 blocks serve the same purpose. Made of hardened or unhardened steel to "tenth" tolerances, their dimensions are 1 × 2 × 3 inches. Usually their weight is reduced by drilled holes, which may or may not be tapped.

Adjustable parallels are precision tools that could be classified as gages or measuring tools. They may be used as precision parallels; however, one of the most important applications of these tools is illustrated in Figure 2-32, Where it is being used to measure the width of a slot. The adjustable parallel is inserted in the opening and expanded to fit the slot. The size of the slot is then measured with a micrometer by measuring over the adjustable parallel. Each adjustable parallel has its own size range; e.g., 3/8 to ½ inch, ½ to 11/16 inch (9.5 to 12.7 mm, 12.7 to 17.5 mm). A set of six parallels that are available commercially will cover all sizes from 3/8 to 2¼ inches, or 9.5 to 57.1 mm.

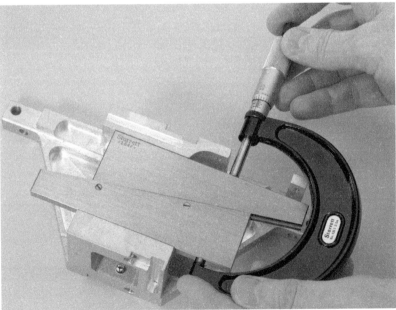

Figure 2-32 Measuring the width of a slot using adjustable parallels and a micrometer.
Courtesy of The L.S. Starrett Company

Going Metric

The metric system adopted by all countries is the International System of Units, or SI. In machine shop practice the application of this system is primarily concerned with linear measurements. The metric units of linear measurement recommended by SI are the meter, kilometer, and millimeter, although the centimeter may also be used in some instances. For all precision machine work the millimeter will be used exclusively. Dimensions less than one millimeter are always expressed as a decimal fraction; common fractions (1/8, 1/4, etc.) are not used as they are with customary inch units.

There are exactly 25.4 millimeters in an inch; thus, one inch is equal to 25.4 millimeters, exactly. Conversely, one millimeter is equal to .039370078 7. . . . inch, which is usually rounded off to .03937 inch. To convert inches and millimeter, from one to another, use the following rules:

multiply inches by 25.4 to obtain millimeters
multiply millimeters by .03937 to obtain inches.

Another method of converting is to use Table 2-3. When the value wanted cannot be found directly in this table, find the values of an equivalent sum; then convert the equivalent values and find their sum to obtain the converted value. This procedure is illustrated in the following examples.

Example 2-2:

Convert .4474 inch to millimeters.
Observe that .4474 = .4 + .04 + .007 + .0004.
Find the millimeter equivalent of each term of the sum and add.

.4	in. =	0.16000 mm
.04	in. =	1.01600 mm
.007	in. =	.17780 mm
.0004	in. =	.01016 mm
.4474	in. =	11.36396 mm

Example 2-3:

Convert 80.92 mm to inches.

80.	mm =	3.14961 in.
.9	mm =	.03543 in.
.02	mm =	.00079 in.
80.92	mm =	3.18583 in.

The converted dimension must be rounded-off to be consistent with the degree of precision required by the job. In Example 2-2, when working to the millimeter equivalent of a "thousandths" of an inch (.001 in.) the answer would he 11.36 mm; when working to "tenths" (.0001 in.) the answer is; 11.364 mm. In Example 2-3 the answer is 3.186 inches and 3.1858 inches when working in "thousandths" and "tenths" respectively. When converting precise dimensions, however, in some cases the converted dimension must not exceed and in others not be less than the original dimension. For example, if the dimension in Example 2-3 is a shaft diameter, the converted dimension may be required not to exceed the original dimension, in which case it would be 3.185 inches when working to "thousandths." If it is a hole diameter the converted dimension may be required not to be less than the original dimension, in which case it should be 3.186 inches as before. In Example 2-2, the shaft and hole diameters would be 11.36 mm and 11.37 mm respectively. This procedure should, however, be applied only when conditions affecting the dimensions are known; otherwise, use the procedure first described.

Table 2-3. Inch-Millimeter and Inch-Centimeter Conversion Table

(Based on 1 inch=25.4 Millimeters, exactly)

INCHES TO MILLIMETERS

in.	mm	in.	mm	in.	mm	in.	mm	in.	mm	in.	mm
10	254.00000	1	25.40000	0.10	2.54000	0.010	.25400	0.0010	.02540	.0001	.00254
20	508.00000	2	50.80000	0.20	5.08000	0.020	.50800	0.0020	.05080	.0002	.00508
30	762.00000	3	76.20000	0.30	7.62000	0.030	.76200	0.0030	.07620	.0003	.00762
40	1016.00000	4	101.60000	0.40	10.16000	0.040	1.01600	0.0040	.10160	.0004	.01016
50	1270.00000	5	127.00000	0.50	12.70000	0.050	1.27000	0.0050	.12700	.0005	.01270
60	1524.00000	6	152.40000	0.60	15.24000	0.060	1.52400	0.0060	.15240	.0006	.01524
70	1778.00000	7	177.80000	0.70	17.78000	0.070	1.77800	0.0070	.17780	.0007	.01778
80	2032.00000	8	203.20000	0.80	20.32000	0.080	2.03200	0.0080	.20320	.0008	.02032
90	2286.00000	9	228.60000	0.90	22.86000	0.090	2.28600	0.0090	.22860	.0009	.02286
100	2540.00000	10	254.00000	1.00	25.40000	0.100	2.54000	0.0100	.25400	.0010	.02540

MILLIMETERS TO INCHES

mm	in.	mm	in.	mm	in.	mm	in.	mm	in.	mm	in.
100	3.93701	10	0.39370	1	0.03937	.1	0.00394	.01	.00039	.010	.00004
200	7.87402	20	0.78740	2	0.07874	.2	0.00787	.02	.00079	.020	.00008
300	11.81102	30	1.18110	3	0.11811	.3	0.01181	.03	.00118	.030	.00012
400	15.74803	40	1.57480	4	0.15748	.4	0.01575	.04	.00157	.040	.00016
500	19.68504	50	1.96850	5	0.19685	.5	0.01969	.05	.00197	.050	.00020
600	23.62205	60	2.36221	6	0.23622	.6	0.02362	.06	.00236	.060	.00024
700	27.55906	70	2.75591	7	0.27559	.7	0.02756	.07	.00276	.070	.00028
800	31.49606	80	3.14961	8	0.31496	.8	0.03150	.08	.00315	.080	.00031
900	35.43307	90	3.54331	9	0.35433	.9	0.03543	.09	.00354	.090	.00035
1000	39.37008	100	3.93701	10	0.39370	1.0	0.03937	.10	.00394	.100	.00039

• For inches to centimeters, shift decimal point in mm column one place to left and read centimeters, thus: 40 in.= 1016 mm = 101.6 cm. For centimeters to inches, shift decimal point of centimeter value one place to right and enter mm column, thus: 70 cm = 700 mm = 27.55906 inches

The Basic Nomenclature of Measurement

The nomenclature of measurement deals with the definitions of the terms encountered in the field of measurement. The topics in the remainder of this book will be more meaningful after a study of these terms.

Nominal Size is the ideal size to which a part, or two mating parts, is to be made. It does not take into account the clearance, or interference necessary in fitting two parts together, so that they will function as desired. Nor does it take into account the deviations in size resulting from the manufacturing process.

Allowance is the intentional clearance between two mating parts so that they will work together as desired when they are assembled. In the case of a press, or shrink fit, allowance is the intentional interference between two parts.

Limits are the maximum and the minimum permissible dimensions of the size of a part. For example, a bore may be specified to have a diameter between 6.750 and 6.756 inches, in which case 6.750 and 6.756 inches are the limits.

Upper Limit specifies the largest size to which a dimension on a part is to be made. In the example above, 6.756 inches is the upper limit.

Lower Limit specifies the smallest size to which a dimension on a part is to be made. In the above example, 6.750 inches is the lower limit.

Tolerance is the specification of the total permissible variation of size of a dimension on a part. The total tolerance in the above example is .006 inch. In some instances this is expressed in terms of plus or minus from a given basic size (basic size is described below). For example, the tolerance in the example above might also be expressed as $\pm$ .003.

Basic Size is that size from which the limits of a part are obtained by the application of tolerances. It is sometimes considered as the "ideal" size of the part. For example, if the ideal size or the basic size of a bore is 6.753 inches and the tolerance is. $\pm$ 003, the limits of the hole would be 6.750 and 6.756 inches.

Unilateral Tolerance refers to a permissible deviation in the basic size in one direction only; i.e., this specification states that the part is to be made with the basic size as the lower limit, or with the basic size as the upper limit. An example of a unilateral tolerance would be 6.750 +.003. Another example would be 6.775 − 002.

continued

Bilateral Tolerance refers to a permissible deviation from the basic size in two directions; i.e., this specification would state that the basic size is neither the upper nor the lower limit and that the part may be slightly larger or slightly smaller than the basic size. Bilateral tolerances may be equal or unequal. An example of an equal bilateral tolerance is 6.753 ±.003 inches. An example of an unequal bilateral tolerance would be

$$6.752 \begin{array}{c} +.002 \\ -.001 \end{array} \text{ inches}$$

Precision refers to the degree of dimensional refinement to which a part is made or to which a machine is capable of working. A part which is made to, or a machine which will produce parts, with a very small dimensional tolerance is said to be precise.

Accuracy refers to the confirmation of a part to the requirements specified by the dimensional tolerance. A part that is within the required dimensional tolerance is accurate. If the required tolerance is very small, say ±.001 inch, the part is also precise. If the tolerance on a machined part is large, say, ± 1/16 inch, a part is accurate if it falls within this range, yet, at the same time, it would be considered to be imprecise.

Tip from a Pro

Using a Micrometer

Text and photos reprinted by permission of
Sandro Di Filippo
and *The Home Shop Machinist*
(May/June 2013).

Because of the preoccupation with precision in the modern machine shop, the micrometer has, not surprisingly, become the preeminent tool for measuring. The idea of precision measuring is not a new one, and many inventors and scientists have spent years trying to develop a simple means of measuring to the tiniest fraction accurately. Even though it is a standard tool in every machine shop, used by countless machinists, the micrometer can be a source of frustration when used incorrectly.

History

The micrometer is a surprisingly old measuring device. The word first appears in the English language in the late 1600s, with some early micrometer screws being made in the 1700s for use in astronomical instruments. A bench micrometer was built by Henry Maudslay in the early 1800s, which he named "The Lord Chancellor" because it was the final word in accuracy in his shop. Remarkably modern looking micrometers were developed in the mid-1800s, and the well-known company of Brown & Sharpe produced its first commercial micrometer in 1867. It has been a mainstay of machine shops ever since.

Design

When you consider how relied upon the micrometer is, it is amazing how simply designed it is. A micrometer consists of only one moving part, and that one part does all the work. If you have a micrometer handy, go get it and take it apart; don't worry, you won't damage it as long as you handle it carefully.

Photo 1 shows a 0-1″ micrometer, with the major parts labelled with the commonly used names. When you unscrew the thimble from the barrel, you end up with two pieces (*Photo 2*). When you have a look inside the barrel or thimble, you will see the heart of the operation of a micrometer, a simple screw (*Photo 3*). This "simple screw" has given ma-

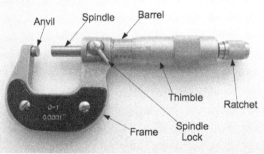
Photo 1

Photo 2

chinists the power to measure to .0001″! The screw, though, does have some special qualities. In inch micrometers, regardless of the diameter of the screw, it is 40 threads per inch (TPI). Simple math tells you why: One turn of the screw equals 1/40 of an inch, which equals .025″. Metric micrometers work the same way, except the pitch of the screw is .5mm.

The spindle lock is self-explanatory. Flicking the lever to the left or right (it varies by manufacturer) will lock the spindle in any position so a size can be referred to after a piece has been measured. Sometimes the lock is a knurled nut (**Photo 4**), but it works in much the same way.

One big problem with using a micrometer is the "feel" – the ability to twist the thimble just enough to make contact with the piece being measured, but not so much that you over-tighten the screw and get an incorrect reading. The ratchet on the end of the thimble, in theory, is supposed to eliminate this problem by controlling the amount of force applied to the screw. Other micrometers are fitted with a feature known as a friction thimble in which a sleeve that slips when a predetermined amount of force is applied to the screw is fit over the end of the thimble.

In my experience, both systems have drawbacks. If the micrometers have a lot of wear and tear and are not clean, the ratchet and friction thimble will no longer work reliably. My micrometers at work don't have a ratchet or friction thimble, but are "plain" micrometers: no lock, ratchet, or friction thimble. The micrometer at the bottom of **Photo 5** is an example of a plain micrometer.

The photo also illustrates another difference between some micrometers, the frame. The upper micrometer has a square frame and the lower one a round frame. Both work in the same way, except square frame micrometers are sometimes more convenient to use when measuring something that isn't round (**Photo 6**).

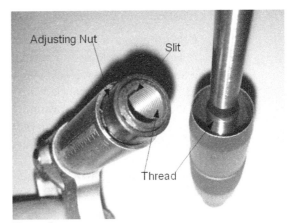

Photo 3

Photo 4

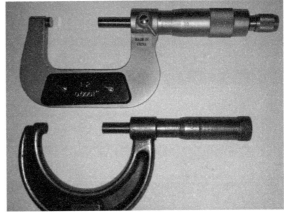

Photo 5

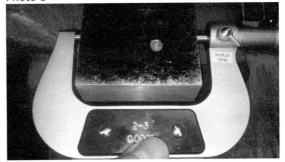

Photo 6

Use

Checking your micrometer before you use it is a good habit to get into. New micrometers are always sold with a checking standard. If the standard is missing, anything of a known size, like a gauge block, can be used. The checking standard, usually just called a standard, is always a nominal size and has the size marked on it. Depending on the manufacturer, they can be straight or round, but are used the same way (***Photo 7***). To check a micrometer, first wipe the anvil, spindle, and both ends of the standard with a clean cloth to remove any dirt or chips. Then, using the appropriate standard, screw the spindle down to it (***Photo 8***). The exceptions are the 0-1″ or 0-25mm micrometers. These can be checked by screwing the spindle down to the anvil. It's important to not over-tighten the spindle. Use light fingertip pressure, allowing your fingertips to slip over the spindle. Also, twist the standard back and forth a few times to be sure it's properly seated. If the micrometer is properly adjusted, it should read "0". If it's out by more than about .0005″ (.012mm), it should be adjusted.

Photo 9 shows two styles of wrenches used to adjust a micrometer. The exact style varies by manufacturer but they all work in the same manner. Turn the micrometer over and on the back of the barrel near the frame you will see a shallow hole. One end of the wrench should fit over the barrel and hook into the hole (***Photo 10***). Twist the barrel until the two zeros line up (***Photo 11***). The barrel will probably be very tight in the frame, so it will take a bit of force to move the barrel. It's not unusual to overshoot the zero mark and have to turn the wrench around and try again. Be patient, it will take a few tries. Keep checking with the standard until everything looks good.

Another adjustment that needs to be made after a lot of use is the play in the micrometer screw. This is one of those adjustments that only needs to be done after many

Photo 7

Photo 9

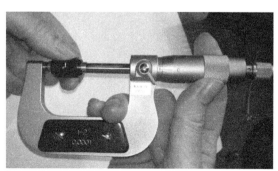

Photo 8

Photo 10

years of use. If you have an older micrometer, you may find it necessary. Have another look at **Photo 3** and you can see the adjusting nut at the end of the barrel. The nut fits over an external tapered thread, and the slits on the inside allow the fit of the thread to be adjusted by turning the nut. The procedure is much like adjusting the barrel, even using the same wrench. It only requires a tiny fraction of a turn to adjust the fit. Once the micrometer has been properly adjusted, it's ready to be used.

Using a micrometer is really very simple. Screw down the spindle until it makes contact with the piece to be measured and read the size off the barrel and thimble. Of course, knowing how to read a micrometer makes it a lot easier to use, so let's work through a few examples to see how it's done.

Micrometers are always sold to measure over a set range. Individual inch micrometers measure 0-1, 1-2, 2-3 inches, etc. Metric micrometers come in 0-25mm, 25-50mm, 50-75mm, etc. The graduations on the barrel of an inch micrometer always start at zero, so you need to read the size off the barrel and add it to the range of the micrometer.

Look at **Photo 12** to see how an inch micrometer works. When you look at the barrel, you will see lines. Each line is worth .025″, and every fourth line equals .100″. The thimble of the micrometer in **Photo 12** has three lengths of lines. The shortest line is worth .0005″, the middle-sized line is worth .001″ (most micrometers only display lines worth .001″), and the longest line is worth .005″. To read the size, just add up the numbers. Start with the largest number on the barrel, in this case .300″. Next, look at the number of lines *not* in contact with the thimble. You can see two lines, which equal .050″. Now read the number shown on the thimble (.022″ is shown here). Add the numbers together: .300 +.050 +.022 =.372″. If the micrometer measured in the range of 1-2″, the final size would be 1.372″. The micrometer in **Photo 12** also has a vernier scale on the barrel for measuring to .0001″.

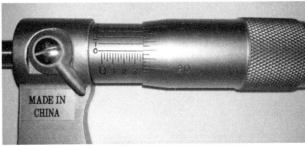
Photo 12

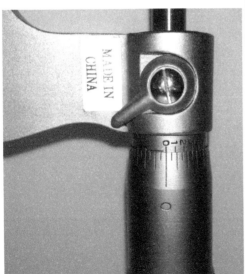
Photo 11

Photo 13

Metric micrometers work in a very similar fashion; look at **Photo 13** to see an example. The numbers on the barrel of metric micrometers start at the minimum size of the micrometer. In this example, the micrometer measures in the range of 25-50mm. Each long line on the bottom of the zero line of the barrel is worth 1mm, each shorter line above is worth .5mm. Each line on the thimble is worth .01mm, with each long line being .05 mm. So just like an inch micrometer, add up the numbers to get the size. In this example the barrel shows 36 mm and the thimble shows .31mm, so the final size equals 36.31mm. Metric micrometers can have a vernier scale that reads to .001mm, but it's not usually necessary, since .01mm is equal to .0004″. Metric micrometers are known to provide a more precise measurement than inch micrometers.

Finally, a few words about using a micrometer to measure something. One problem with micrometers is the pitch of the screw; it's so fine that you can very easily over tighten it and get a false reading. It can take some practice to develop the right feel to know when the spindle has made contact with the part being measured. Slipping the micrometer over the part while gently rocking it from side to side can be helpful. One way to get comfortable using a micrometer is to practice measuring something that's a known size (like the shank of an end mill) and see how accurate you can get.

Micrometers are available up to any size, depending on how much you want to pay. I have seen them as big as 40″. Most hobbyists can make do with just two, a 0-1″ and 1-2″ (0-25, 25-50mm). Add bigger ones as you need them.

3 Machine Shop Tools and Materials

The measuring tools covered in Chapter 2 will help make any projects you tackle a success. But you will need other tools. The big players in the machine shop tool chest, such as drill presses, lathes, and mills, are covered in chapters of their own. But there is another group of hand and power tools that will make the work go easier and quicker. Even if you are setting up your first machine shop, you may already own some of the tools in this chapter.

You will also need to know about the metals you will be working with in your shop. There is a wide variety available. This chapter will help you pick the materials that are best for your project.

Scribers

When you are setting up to cut or shape a piece of metal, you will need layout lines to guide the tool you will be using. Scribers are hardened, pointed tools that allow the machinist to scratch lines in the metal, usually by following some sort of guide, such as a straight edge. See Figure 3-1. Where a pencil or felt-tip marker may smear, a scribed line is etched into the metal.

An alternative to scratching the metal surface is to coat the workpiece with a layout dye and scratch the layout lines in the dye. See Figure 3-2. This helps you avoid scratching the work, and in the case of some metals, such as galvanized steel, scratching the surface could expose the metal to corrosion. Dyes can be brushed or sprayed on. Clean off the dye using a solvent recommended by the manufacturer.

Figure 3-1 Scribers allow the machinist to etch layout lines directly on the metal.
Photo courtesy of Tom Lipton

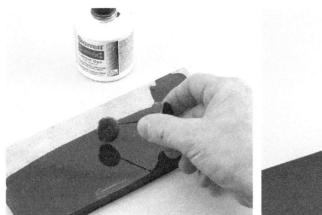 

Figure 3-2 Layout dye prevents etching the metal but still provides a guide for machining.
Dyes can be brushed or sprayed on.
Courtesy of The L.S. Starrett Company

Punches

Punches help the machinist locate the center of a hole. There are two types of punches: prick punches and center punches. They are often used together on the same job. Because of its sharper point, use a prick punch to find the center of the hole. Then follow with the center punch. See Figure 3-3. The wider head enlarges the starter hole somewhat and provides a good "seat" for the drill that will follow. Use a hammer to drive the tip of the punch home. Automatic punches do away with the need for a hammer, which means it is easier for the machinist to see the layout lines when using the punch.

Figure 3-3 Here is an example of a center punch.
Photo courtesy of Tom Lipton

Files

A set of files of different sizes and shapes is indispensible to the home machinist. Use them to sharpen tools, smooth out rough surfaces, and clean up edges, among other uses. The main types of files are single-cut files, which have rows of parallel teeth across the surface of the file. See Figure 3-4. These files produce smooth finishes. Double-cut files have crisscross rows of cutting teeth. These files produce a coarse surface. Files are available in a number of shapes and sizes. Match the shape to the workpiece for best results.

The file teeth can become clogged with debris, called pinning. Keep the tool clean by brushing with a file card, a brush with short steel teeth. Store files in a wall rack or wrapped in paper to prevent damage from other tools.

Figure 3-4 Here is a single-cut file smoothing a part.
Photo courtesy of Tom Lipton

Filing Technique

Be sure the piece to be filed is stable by clamping it to a worktable or placing it in a vise. To remove a lot of material, apply downward pressure as you push the file forward. Only apply pressure on the forward stroke; lift the file clear of the work on the return. The coarseness of the file and the amount of pressure applied will determine how much material is removed.

To smooth a surface, use the draw filing technique by holding the file perpendicular to the direction of the stroke. Apply even pressure on both the forward and the return strokes to achieve a smooth surface. As when using sandpaper on wood, start with a coarse file and move to smoother ones as the job progresses.

Figure 3-5 A good bench vise is like having an extra set of hands in the shop.
Photo courtesy of Tom Lipton

Bench Vises

If you never had a bench vise in your shop, and then decided to add one, you will wonder how you ever got along without it. A good vise is an extra set of hand—very firm-gripping hands—that holds work while you file, saw, drill, or polish it.

Vises come in a variety of sizes and configurations, including vises that have swivel bases and attached anvils. For the home machinist, a heavy duty vise with 3 ½- to 4-inch jaws that can be bolted to a workbench should suffice. See Figure 3-5. For smaller shops, you can find smaller tools or you can opt for a portable vise. Some models attach to a workbench with a vacuum seal while others clamp to the end of the bench. When not in use, you can stow them out of the way.

To protect delicate work, use jaw liners. You can buy magnetic liners, or fabricate your own out of some soft material, such as copper or even wood.

Drill Vises

Drill or mill vises are designed to be used with a mill or drill press and are often sold as accessories by tool manufacturers. See Figure 3-6. The jaws are often smooth. Most can be attached directly to the mill table for drilling operations.

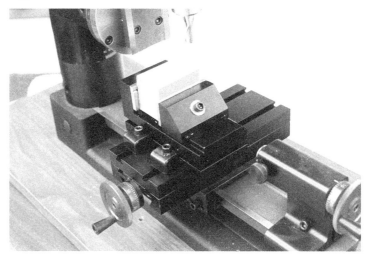

Figure 3-6 Drill, or mill, vises have smooth jaws and are often sold as accessories by mill manufacturers.
Photo courtesy of Sherline Products, Inc.

Clamping Rounds

When working on a large diameter round material, clamp it below the vise jaws as shown in Figure 3-7. This will ensure that the work is held steady. For thinwalled tubes, insert a wooden dowel to help prevent damage to the sides of the tube.

Figure 3-7 Clamp round work below the tops of the jaws to hold it steady.
Photo courtesy of Tom Lipton

Saws

As discussed later in this book, precise cutting and machining of metal are usually carried out on a lathe or mill. But often the home machinist will need to cut large sections of bar stock or sheet metal down to a manageable size for machining. In those cases, a metal cutting saw will be used.

Hacksaws

Use a hacksaw to cut large sections of metal down to size. The replaceable blades are 10 to 12 inches in length with 14 to 32 teeth per inch of blade. Select blades based on the thickness of the material to be cut. In general, thick material requires fewer teeth per inch than thinner material. See Table 3-1. The recommendations in the table are from Diamond Saw Works.

Table 3-1 Selecting Hacksaw Blades

Teeth Per Inch	Material Thickness
14 TPI	5/8 in. and over
18 TPI	¼ in. to 5/8 in.
24 TPI	1/8 in. to ¼ in.
32 TPI	5/16 in. and under

Figure 3-8 Use a hacksaw to cut metal down to a manageable size before machining.
Photo courtesy of Tom Lipton.

Hacksaw blades are made of carbon steel, high speed steel, or carbide. If possible, keep two hacksaws in the shop—one with 14 teeth per inch and the other with 32 teeth. See Figure 3-8. This setup will take care of most of the cuts you will perform, and having two saws handy will save the time it takes to switch out the blades.

Other Saws

Some of the other saws that may be useful in the home shop include reciprocating saws (Sawzall is a common trade name). These tools will cut through metal, but they are not very good at precise, close tolerance work. Circular saws and chop saws usually used to cut wood can go through some types of metal as well. See Figure 3-10.

Metal cutting band saws can cut off stock and cut contours in metal. See Figure 3-11. Band saws contain a long metal blade that circulates on two wheels. The blade passes through the adjustable work table. It probably is not the first tool the home machinist should buy, but they are a versatile tool that does have many applications, especially in the hands of an experienced user. Unlike wood-cutting band saws, metal cutting models operate at a much slower blade speed, usually in the 40 to 350 feet per minute range, depending on the material and the material thickness.

Using a Hacksaw

When inserting a new blade, make sure the teeth face away from the handle. The blade cuts on the forward stroke. Secure the work in a vise and then scribe a notch to start the cut. Use your thumb as a guide as in Figure 3-9. Once started, hold the saw with both hands—one on the handle and the other on the front of the tool. Use long, steady forward strokes to make the cut. Make the cut as close to the vise as possible.

When cutting sheet metal, clamp the sheet to a piece of wood to keep the metal from bending. Cut through both the metal and the wood.

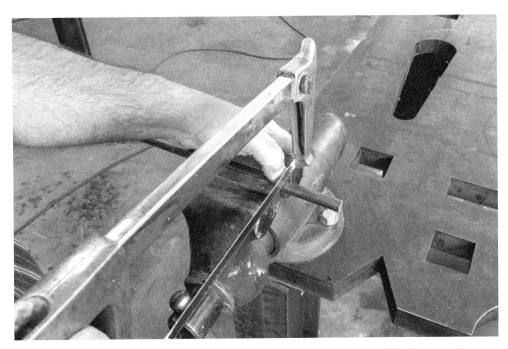

Figure 3-9 Use your thumb as a guide to create a starter notch in the metal.
Photo courtesy of Tom Lipton

Figure 3-10 Some circular saws can handle rough cutting in the shop with the right blade.
Photo courtesy of Tom Lipton

49

Bench-Top Grinders

Bench-top pedestal grinding machines are used primarily to sharpen tools in the home workshop. Outfitted with the right type of wheel, they can also aid in finishing and polishing work, as well as removing rust from metal. The tools have two wheels, which allow the user to keep two different types of abrasive wheels mounted at all times.

Use a Push Stick

As with any power saw, use a push stick to keep hands away from the moving blade. The setup shown in Figure 3-11 uses a section of scrap lumber that the operator braces against his stomach as he guides the metal into the band saw blade.

Figure 3-11 A push stick, or in this case a "belly board," keeps the operator's hands away from the moving blade.
Photo courtesy of Tom Lipton

Figure 3-12 Bench-top grinders have two wheels and can be used to sharpen tools or polish finished work among other uses.

Abrasives. The abrasive materials used on grinding wheels are either aluminum oxide, silicon carbide, or diamond. The type of abrasive should be matched to the material being ground and type of work being done. For example, a carbide abrasive should be used when sharpening carbide tools. A list of abrasive grain sizes is shown in Table 3-2.

Table 3-2 Grinding Machine Abrasives

Grain Size	Grit Number
Coarse	8-24
Medium	30-60
Fine	70-180
Very Fine	220-600

Grade. Grinding wheels are termed either hard or soft, which is determined by the bonding method that holds the grinding material in place. The grade or hardness markings may seem counterintuitive. Hard grade wheels are used on soft metals because the bonding material holds the grit in place longer. Soft grade wheels are used on hard metals because the abrasive material breaks away faster, exposing new sharper abrasive material. The grade of a wheel is listed as a letter of the alphabet—"A" is the softest grade; "Z" is the hardest.

Truing and Dressing. Truing an abrasive wheel means removing material from the perimeter of the wheel so that it is again round, see Figure 3-13. Dressing means removing clogged abrasive material from the wheel so that new sharper abrasive material is exposed. There are a variety of tools available to accomplish both tasks.

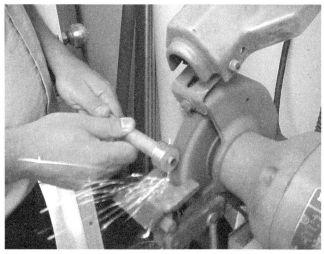

Figure 3-13 Here is an example of dressing the wheel on a pedestal grinder.
Photo courtesy of James Harvey

Finishing. Bench-top grinders can also be used to remove burrs and other imperfections from parts or to give a part a final polishing. For polishing, attach a buffing wheel to the grinder. The wheel is used with a polishing compound that you apply to the wheel. Use only one type compound for each buffing wheel. As when sanding wood, start with a coarse compound and work up to finer and finer grits until you achieve the finish you want. When polishing, wear gloves and hold the work in both hands on the lower front portion of the wheel.

Machine Shop Materials

A machinist working in an industrial machine shop handles numerous metals in the course of a day. Each of those metals has specific characteristics regarding strength, workability, resistance to corrosion, etc. As a professional, the machinist must be able to identify metals and choose the correct one to complete a project. The home machinist has the same responsibilities, but to a lesser degree. Knowing which metal or metals to choose for a project helps ensures the project's success.

Types of Metals

Metals are classified as either ferrous or nonferrous. Ferrous metals are composed mostly of iron. Steel is a ferrous metal. These metals are the hardest and strongest metals you will use. Nonferrous metals do not contain iron as a base metal—aluminum, copper, and nickel are nonferrous metals. These products are available in a variety of shapes and sizes. See Figure 3-14 and 3-15.

Types of Steel. Steel is a combination of iron and carbon, and often other elements.

Figure 3-14 Store flats on edge to save space. The owner of this shop used a color coding system to identify materials.
Photo courtesy of Tom Lipton

The combination of elements and the steel manufacturing process determine the characteristics of the steel, such as its workability, strength, hardness and the like. In general, steel products are referred to as carbon steels and alloy steels.

Carbon Steels. These products are classified as low carbon, medium carbon, and high-carbon steels. The more carbon the material contains, the stronger the steel. Carbon steels contain between .3 and 1.7 percent carbon. This is also listed as points, which are hundredths of 1 percent. So, steel with .30 carbon is 30 points. Here are the principal types of carbon steels:

- **Low-Carbon Steel.** Less than .3 percent carbon. This is often called mild steel. This steel is relatively soft and often used for parts that do not need to be hardened.

- **Medium-Carbon Steel.** Between .3 and .6 percent carbon. Stronger than low-carbon steel. Used for tools like screwdrivers and wrenches.

- **High-Carbon Steel.** Between .6 and 1.7 percent carbon. Also called tool steel. Used for cutting tools and drills.

- **High-Speed Steel.** Contains between .85 and 1.5 percent carbon. This is the steel used for lathe and mill cutting tools.

Figure 3-15 Some metals are available in block form.
Photo courtesy of James Harvey

Alloy Steels. In order to obtain the variety of characteristics required of steel products, manufacturers add alloying elements. Each of the elements contributes a specific characteristic to the finished steel product. See Table 3-3 for a sampling of alloying elements.

Table 3-3   Alloying Elements

Alloying Element	Enhanced Characteristics
Carbon	Surface hardness, strength
Chromium	Resistance to corrosion
Lead	Improves machinability
Nickel	Increases strength and resistance to abrasion
Molybdenum	Toughness and resistance to shock
Tungsten	Used with cobalt in tool steels

Numbering System

To help identify the components that go into the hundreds of steel products, the industry has developed numbering systems that tell the purchaser what the steel is composed of and, therefore, the characteristics of the metal under certain conditions. A system developed by the American Iron and Steel Institute (AISI) and the Society of Automotive Engineers (SAE) is the most widely used. Each product is assigned a four-digit number. The first number is the principal alloying element—1 is for carbon, 2 is for nickel, 3 is for nickel-chromium, etc. The second digit shows the percentage of the principal alloying element. The last two digits provide the approximate percentage of carbon in the steel. So the product labeled 1045 is a carbon steel with no other alloy present and .45 percent carbon. An "L" inserted between the second and third digit means that lead was added to improve machinability. There is a similar numbering system for aluminum.

To get started selecting metals for your project, visit the Website of a steel supplier. For example, the site onlinemetals.com has information on a variety of products. Some popular materials include 1018, 12L14, 4130 for steel parts, and 6061 for aluminum. When selecting metal for a project, remember to keep not only the characteristics of the material in mind, but also the shape and size the metal is available in.

Other Terms for Metal Selection

Here are some other terms you might hear when talking "shop" at a metal supplier. Most of them are more relevant to industrial machining, but may help you when picking material for a project.

Hardness Tests. The industry uses two main testing procedures to determine the hardness of a metal: The Rockwell Hardness Test and the Brinell Hardness Test. In both tests a metal object is dropped onto a test sample. The Rockwell test measures the depth of the impression; the Brinell measures the area of the impression.

Heat Treatment. There are a number of ways to treat metals.
Hardening. This is metal that is heated to a critical temperature and then cooled by plunging the material into water or oil or allowed to air cool. The process in creases the hardness but also makes the material more brittle caused by the hardening process.
Annealing. This is metal that is heated and then cooled slowly to a temperature below the "critical temperature." The metal becomes softer but more ductile.
Tempering. This step reduces brittleness. The metal is heated slowly and allowed to cool.

Hot-Rolled and Cold-Rolled Metals. Two metal manufacturing processes that determine the qualities of the final product. Rolling a material breaks up its internal structure. Hot rolling is done under heat while cold rolling is accomplished at room temperature. Hot-rolled metals are less likely to warp.

Tip from a Pro

Basic Layout Skills

Text and photos reprinted by permission of
Sandro Di Filippo
and *The Home Shop Machinist*
(September/October 2012).

When I first entered this trade decades ago, I would spend hours at my bench scribing lines, circles, and arcs. Layout was something that all machinists and toolmakers did every day throughout their career, but now is almost never done since CNCs and DROs have made it obsolete.

But for the amateur working with limited equipment, it is still a useful skill. The basic tools – a combination square, a scriber, a center punch, and a hammer – are simple to use and the skills needed are basic and easily learned. With some proper lighting and layout dye good results can be achieved right away. With experience, incredibly complex and accurate layouts will become possible. To illustrate this last point, let me point out that until the invention of the milling machine and the dividing head in the mid to late 1800s gear teeth were filed by hand to a layout line. Let's start by having a look at the basic tools required.

I have done many layouts without using any layout dye, but you should apply some kind of marking fluid to make the lines easier to see. Layout dye is a thin fluid that is brushed or sprayed on; it's usually blue in color, but can be green or red. Ensure that there is plenty of ventilation, since its odor can be pretty strong. Also, use care as it will produce a deep and lasting stain on whatever it touches, especially clothes and skin. My own preference is to use a marker *(Photo 1)*, ever since I spilled a bottle of dye on the floor of my basement shop a few years ago. Whichever product you choose, apply it to the workpiece and let it dry for a few minutes.

To make and see lines, a scriber and magnifier are indispensable *(Photo 2)*. Get a good scriber; don't try to get by with something like a sharpened nail (I've seen it, seriously). I like the type with a reversible point; it prevents me from getting jabbed in the thumb when I'm searching through a drawer for it. Another style is the double ended scriber with a straight point and an angled point. If your eyes are like mine, a magnifying lens is useful when center punching layouts.

This leads to two more essential tools: the center punch and a small hammer. Note that I said a small hammer, even smaller than the one seen in *Photo 3*. Something like a tack hammer might be good to start with since the initial center punch marks need to be very light and a heavy hammer can make this difficult to achieve. Starrett makes a small toolmaker's hammer (ID No. 815) with a built in magnifying lens, but at $70 it's a little too pricey for my needs.

Finally, we come to the combination square, also known as a machinist's square, made up of four pieces: the scale, the square head, the center head, and the protractor *(Photo 4)*. It has been used by machinists since it was first invented by L.S. Starrett in the 1870s to lay out lines perpendicular to or at an angle to any edge. The length of the scale is normally 12″, but scales as short as 4″ and as long as 24″ are available, with graduations ranging from an 1/8 inch down to 1/50 inch, as well as metric.

Photo 1

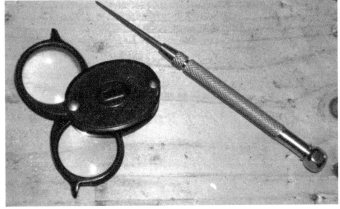

Photo 2

Photo 3

The square head has two built-in angles, 90° and 45°, and in its most basic function can be used to check angles either by itself *(Photo 5)* or with the scale inserted *(Photo 6)*. For a more advanced function, using care, the scale can be used to lay out lines within .010″ or better. The protractor can be employed to measure *(Photo 7)* and lay out any angle *(Photo 8)* with a precision of plus or minus 1°. For the purpose of this article, I won't be covering the center head but it is a useful device for finding the center of a round or square bar.

Photo 4

Photo 5

Photo 6

Photo 7

Photo 8

Photo 9

One final necessity is good lighting, as shown in **Photo 9** where you can see my setup. I have the part that needs a layout in the vise with a gooseneck lamp attached to the wall right behind my bench, giving me plenty of light right where I need it. Remember, if you can't see anything, you can't do anything.

Layouts are always done from only two edges and the edges should be finished (**Figure 1**). Even if the holes, or any other feature, are laid out symmetrically on the part, everything must be done from only two reference edges (usually referred to as datums). Doing otherwise may allow minute errors to appear due to the overall width or length of the part being oversize or undersize or out of square. The two reference edges must always be square to each other.

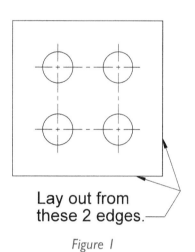

Lay out from these 2 edges.

Figure 1

Photo 10

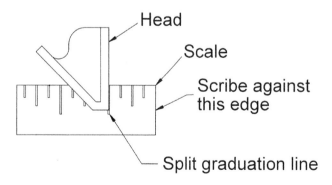

Figure 2

Here is the method to lay out a hole 1″ from the edge of a workpiece: To start, set the scale so the 1″ graduated line is split by the head *(Photo 10 and Figure 2)*. Lay the head against the part and scribe a line along the end of the scale; first from one edge, then from the second edge *(Photos 11 and 12)*. Once the lines are in the right place, a center punch is used to mark the point where they intersect *(Photo 13)*.

Line up the center punch by sighting along one line then, without moving the center punch, look down the other line. *Figure 3 and Photo 14* show the general technique. Take your time at this stage because it is very easy for the punch to move slightly as you shift your view from one line to the next.

Photo 11

Photo 12

Photo 13

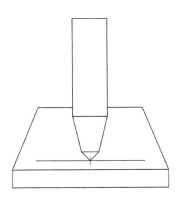

Figure 3

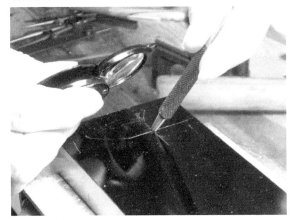

Photo 14

You may or may not need a magnifier to see, depending on your eyes and the lighting. When it looks good, hold the center punch vertical and lightly mark the intersection point *(Photo 15)*. The weight of the hammer should be all the force needed for the initial mark. Double-check the punch mark by sighting along both lines again. If the punch mark is off, lay the center punch over slightly and tap the point in the direction you need to go, then recheck the punch mark *(Photo 16)*. When all looks good, carefully deepen the mark, being sure to hold the punch vertical. *Photo 17* shows the final result.

To be honest, it will probably take you longer to read this article than to do your first layout. Just take your time and double-check everything before you start drilling. Remember, practice makes perfect, so head out to the shop and start doing some layouts.

Photo 15

Photo 16

Photo 17

Tip from a Pro

Using a Bench Grinder

Text and photos reprinted by permission of
Sandro Di Filippo
and *The Home Shop Machinist*
(July/August 2014).

Bench grinders typically available for the home workshop are 6″, 8″, or 10″. The number designates the diameter of the wheel. The larger size, while more expensive, tends to be the better choice. As a general rule, the larger the diameter of the wheel, the cooler and quicker the wheel will cut. A larger wheel has more surface area and has more time to cool down. Also, the larger surface area provides more sharp particles to cut with. I compromised and use an 8″ grinder in my shop *(Photo 1)*. For convenience, I mounted my grinder on a homemade pedestal, technically making it a pedestal grinder. Pedestals can be bought, and I recommend them since they more easily allow you to move the grinder around as required.

Safety

While I expect every reader of this magazine uses safety glasses, it's important enough to be repeated. Always wear safety glasses, or a face shield, or both, when using a grinder. Many times in my career I've had a hot spark bounce off the lenses of my safety glasses, and a few have even embedded themselves.

Many readers will notice that the clear plastic guards mounted over the grinding wheels are missing on my grinder. For readers who don't know what I'm talking about, many machines are sold with a clear guard positioned over the wheel to prevent sparks from flying up into the operator's face. Every shop I've ever worked in has removed these guards. While I'm not recommending this practice, there is a good reason for it. After some use, the plastic gets scratched, making it difficult to see through. I like to know where my fingers are at all times because the wheel can easily remove skin.

Photo 1 An eight-inch bench grinder mounted on a pedestal.

This brings up another safety concern: fingers, and where they should be. I still have a scar from a grinder related injury I got several decades ago. Moving the piece from side to side, you can slip off the wheel, catching the knuckle of the first finger *(Photo 2)*. This type of injury is more common when doing heavy removal with a grinder. Move your hands with deliberate care when using a grinder, especially if you have little experience with one, since grinding wheel injuries are very painful and take a long time to heal because the wound is so jagged.

Mounting the Wheel

When you get a new wheel, before mounting it, check that it isn't cracked. Perform a visual check first and then a check known as "ringing" the wheel. Let the wheel balance on a finger through the mounting hole. Gently tap the wheel with something non-metallic, like a piece of wood or plastic. I usually use the handle of a hammer or the plastic handle of a screwdriver *(Photo 3)*. When struck, you should hear a clear ringing sound, like a bell. If you hear a dull thud, the wheel has some flaw and should not be used. If a flawed wheel is mounted on a grinder, when the wheel gets up to speed, centrifugal force will cause the wheel to explode. Having had wheels explode on me in the past, it's not an experience I'd recommend you try. Although it's unlikely that bench grinders present the same level of danger that industrial grinders offer, I don't think it's a good idea to be surrounded by what is essentially shrapnel.

Another cause of a damaged wheel is improperly mounting them. In **Photo 3** you will see a piece of paper in the center of the wheel that serves an important function. It's called a blotter and is primarily

Photo 2 The knuckle of the first finger can be easily injured.

Photo 3 Tap the wheel with plastic or wood to check for flaws.

there to take up any unevenness between the side of the wheel and the clamping plates. It acts as a cushion; otherwise, the plates would only make contact on the high spots when tightened. The plates would crush down the high spots on the side of the wheel, possibly causing a fracture. It's not a good idea to re-use blotters, though it is frequently done. After repeated use, the blotter becomes flattened and loses its cushioning ability. Blotters are available separately from industrial suppliers. If you need one, go get it, it could save your life.

Photo 4 Checking clearance with a pencil.

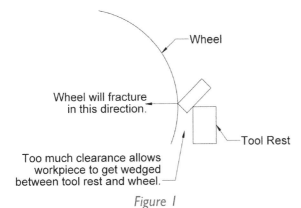

Wheel

Wheel will fracture in this direction.

Tool Rest

Too much clearance allows workpiece to get wedged between tool rest and wheel.

Figure I

Photo 5 A diamond cluster makes an excellent dresser for grinding wheels.

Never operate a grinder without the wheel guards in place. Should a wheel explode, the only thing protecting you will be the guards. The wheel guard will contain the wheel and prevent flying shrapnel from injuring you.

Another important safety concern is clearance between the tool rest and the wheel. There should only be a small gap between the wheel and the rest. A simple check to see if you have too much clearance is using a pencil to measure the gap *(Photo 4)*. If the pencil fits between the wheel and the tool rest, you have too much clearance. Too much clearance can allow a finger to get pulled into the wheel – wheels tend to be "grabby."

A similar problem can occur when grinding a workpiece. Excess clearance can allow the workpiece to become wedged between the wheel and the tool rest, splitting the wheel. Centrifugal force takes over after that, leading to a complete split and an exploding wheel *(Figure 1)*.

Grinding Wheels

The most commonly used wheel, even in commercial shops, is still the aluminum oxide wheel. It is characterized by a grey color and is a good general-purpose grinding wheel. It can be used for everyday sharpening and also for material removal. Another type of wheel is the silicon carbide wheel, usually used for grinding carbide tooling. Silicon carbide wheels are green in color and are quite soft. They tend to require frequent dressing and are primarily used for carbide grinding.

Other types of wheels that are available are ceramic, Borazon, and diamond. These wheels are only required for specialized applications and likely wouldn't be found in the amateur's shop.

Photo 6 A homemade clamping washer to replace the stamped original.

Dressing the Wheel

The time honored way of dressing a wheel is to use a contraption known as a Huntington dresser, though I have never heard anyone call it that. The Huntington dresser consists of a series of heat-treated serrated cutters and wheels. The method is to crush the surface of the wheel and true and clean the wheel. I've heard it referred to crush dressing, but I'm not sure if that is the correct term. I've never liked using one and years ago adopted a diamond cluster, originally from a diamond core drill, for dressing a grinding wheel *(Photo 5)*. If you search through eBay, you can find one that isn't too expensive.

Vibration

When using your grinder, there should be little vibration felt. If your bench top is shaking violently or the grinder and pedestal start to walk across the floor, something is wrong. In my experience, there are three probable areas to check when vibration occurs. When I first used my grinder, it vibrated badly. I checked the screws that attach the base to the motor and found two of the four screws were loose. In addition, I replaced the inner stamped washers that clamp the wheel to the spindle with a pair I machined from aluminum *(Photo 6)*. I made the washers a very close fit on the shaft and the faces square to the bore. These two fixes eliminated all vibration from my grinder. The other common source of vibration is a bad wheel. If you find that your grinder vibrates badly, try replacing one or both grinding wheels to try and eliminate it.

4   Drill Presses

Drill presses open, enlarge, and finish holes by drilling, reaming, boring, counterboring, countersinking, and tapping. These operations can also be performed on other machines, such as lathes and mills. However, drill presses have a large and easily adjustable operating range, and they are often much quicker to set up that other machine tools. As with other machine shop tools, drill presses are available in a variety of sizes and designs. The horsepower of drill presses may range from a fractional horsepower for drilling small holes to as much as 50 horsepower for driving a heavy-duty industrial, multiple-spindle drill head that is capable of drilling many holes simultaneously.

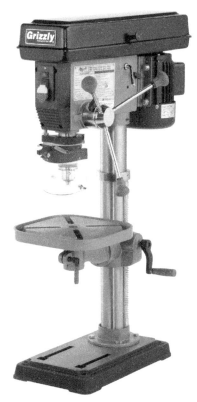

Figure 4-1 This is a typical bench-top drill press for home shop use.
Photo courtesy of Grizzly Industrial, Inc.

Drill Press Basics

Upright drill presses are general purpose drilling machines. They consist of a base; a column, which supports an adjustable table; and the drill head and its components. Benchtop models, such as shown in Figure 4-1, have the same basic configuration, except that the column is much shorter. The adjustable table provides a convenient method of aligning workpieces with the drill. The spindle housing may be raised or lowered, and it contains the feed mechanism. Some newer models are equipped with a laser to help the operator align the work with the spindle. Most machines are capable of a variety of drill speeds. The operator changes speeds by manipulating the belts of a stepped pulley system, or, on some models, the speed can be changed by simply rotating a dial on the motor. A depth stop allows the operator to drill holes to exact depths.

Twist Drills

The basic metal-cutting tool for opening holes in metal is the twist drill. Its apparent simplicity and the ease with which it can be used tend to obscure the fact that it is a complicated cutting tool with a very complex cutting action. See Figure 4-2. The twist drill must withstand high torque and thrust forces, as well as the temperatures resulting from the formation of the chip. Twist drill sizes are given in fractions of an inch, number sizes, and letter sizes; metric drills are specified in terms of a millimeter. Holes produced by twist drills are usually somewhat oversize, as shown in Table 4-1.

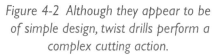

Figure 4-2 Although they appear to be of simple design, twist drills perform a complex cutting action.

Table 4-1. Oversize Amount Normally Cut by a Drill Under
Normal Shop Conditions, in Inches

Drill Dia, Inch	Amount Oversize, Inch		
	Average Max	Mean	Average Min
1/16	.002	.0015	.001
1/8	.0045	.003	.001
1/4	.0065	.004	.0025
1/2	.008	.005	.003
3/4	.008	.005	.003
1	.009	.007	.004

Courtesy of the Metal Cutting Tool Institute

Tool Tip

The accuracy of the drilled hole is dependent on the following factors:

1. The actual size of the drill diameter
2. The accuracy of the drill point
3. The accuracy of the drill chuck and drill sleeve
4. The accuracy and rigidity of the drilling machine spindle
5. The rigidity of the drilling machine
6. The rigidity of the workpiece and the setup.

Sometimes slightly rounding the corners of the lips of the drill at the margins with a hone will improve the accuracy of the drilled hole. Since the tolerance of most drilled holes is usually quite liberal and the holes produced by twist drills are usually somewhat oversize, it is possible to drill a rather wide range of metric size holes with inch size twist drills.

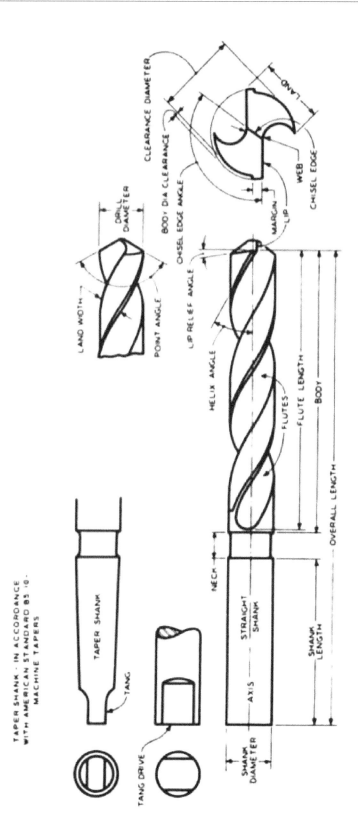

Fig. 4-3 Nomenclature of a twist drill.

Twist Drill Geometry

The flutes of most twist drills have a standard helix angle and these drills can be used to drill almost all materials. See Figure 4-3. For general purpose work, twist drills having a non-standard helix angle are not required; however, in cases where certain materials must frequently be drilled, non-standard helix angle drills may have an advantage. High helix angle twist drills are made and they are recommended for drilling low tensile strength materials such as aluminum, magnesium, die casting alloys, and some plastics. There are also low helix angle twist drills which are used to drill soft bronze and sheet metal. These materials are also drilled using straight fluted drills, having a zero degree helix angle, because they will not tend to pull or run ahead of the feed, and when breaking through the hole, they will have less tendency to grab. Twist drills having a low helix combined with a thicker web are used when drilling some very hard and tough materials.

After a twist drill has been made, its geometry is fixed except at the point, where it may be ground to meet different requirements. The point geometry consists of the point angle, lip relief angle, chisel edge angle, the lip lengths, and the chisel edge length. Suggested lip relief angles are given in Table 4-2. The lip relief angle is dependent on the size of the drill and on the type of material being drilled.

Table 4-2. Suggested Lip Relief Angles at the Periphery

Drill Sizes	Drill Diameter Range		Suggested Lip Relief Angle at Periphery		
	Inch	Mm	For General Purpose Drills	Hard and Tough Materials	Soft and Free Machining Materials
No. 80 to No. 61	.0135 to .0390	0.35 to 1.00	24°	20°	26°
No. 60 to No. 41	.0400 to .0960	1.05 to 2.45	21°	18°	24°
No. 40 to No. 31	.0980 to .1200	2.50 to 3.05	18°	16°	22°
No. 30 ¼″	.1285 to .2500	325 to 6.35	16°	14°	20°
F to 11/32	.2570 to .3433	6.55 to 8.75	14°	12°	18°
S to ½″	.3480 to .5000	8.85 to 12.70	12°	10°	16°
33/64″ to 3/4″	.5156 to .7500	13.10 to 19.05	10°	8°	14°
49/64″ and larger	.7656-	19.45-	8°	7°	12°

Drill Press Operation

Everyone has their own way of doing things. But when it comes machine shop practice, there are a few basic rules to follow. They are:

- Plan how to do the job.
- Gather the tools and materials you will need.
- Prep the workpiece and the machine for the job.
- Start the work, proceed with care and thought, complete the project.

Drilling Speeds

For any machining operation, including drilling, the cutting speed is generally given in terms of feet per minute (fpm). It describes the rate of movement of a cutting tool relative to the workpiece. For example, a cutting speed of 100 fpm would describe a cutting tool traveling in a straight line at a uniform rate of motion, or speed, such that the tool would travel a distance of 100 feet in one minute. The effect would be the same if the tool was stationary and the workpiece moved past the tool at this speed. In drilling, the tool moves, but it rotates instead of moving in a straight line. This is the turning speed and is measured in revolutions per minute (rpm).

The recommended cutting speeds for drilling and reaming different materials with high-speed steel drills are given in Tables 4-3 through 4-7 in the columns labeled "Drilling" and "Reaming." Use this information to determine the turning speed, see "Figuring Drill Turning Speeds." The tables were taken from industrial sources, and may be more information than the home hobbyist needs. But on the other hand, you never know when you will come across a material that you are unfamiliar with. In those cases, these tables will come in handy.

The cutting speed is dependent on the material from which the drill is made, the material from which the workpiece is made, the hardness of the work material and its heat treatment. Another factor affecting the cutting speed is the length of time desired during which the tool will perform satisfactorily, called "tool life." Generally, a cutting tool operating at a lower cutting speed will last longer than when operating at a higher cutting speed. Twist drills are made from several types of high speed steel; and less expensive drills are made from carbon tool steel. Carbon tool steel drills can only be operated at one-half of the cutting speeds listed in these tables.

Parts of a Twist Drill

Axis—The imaginary straight line forming the longitudinal center of the drill.

Body—That portion of the drill extending from the shank, or neck, to the outer corners of the cutting edge.

Flutes—The helical grooves cut, or formed, in the body of the drill to provide cutting lips, to permit removal of chips, and to allow cutting fluid to reach the cutting lips.

Land—The peripheral portion of the body between adjacent flutes.

Body Diameter Clearance—That portion of the land that has been cut away so that it will not rub against the walls of the hole.

Margin—The cylindrical portion of the land which is not cut away to provide clearance.

Web—The central portion of the body that joins the lands. The extreme ends of the web form the chisel edge. The thickness of the web is not uniform, but increases from the point toward the shank.

Point—The cutting end of the drill, made up of the ends of the lands and the web. In form it resembles a cone, but departs from a true cone to furnish relief behind the cutting lips in order that they can penetrate into the metal and form a chip.

Point Angle—The angle included between the cutting lips, projected upon a plane parallel to the drill axis, and parallel also to the two cutting lips.

Lips—The two cutting edges extending from the chisel edge to the periphery.

Relief—The result of the removal of tool material behind or adjacent to the cutting lip and leading edge of the land, to provide clearance; prevent heel drag; and to allow the cutting lips to penetrate into the work and form the chip.

Lip Relief Angle—The axial relief angle at the outer corner of the lip. It is measured measured by projection into a plane, tangent to the periphery, at the outer corner of the lip.

Chisel Edge—The edge at the end of the web that connects the cutting lips.

Chisel Edge Angle—The angle included between the chisel edge and the cutting lip, as viewed from the end of the drill.

Helix Angle—The angle made by the leading edge of the land with a plane containing the axis of the drill. This angle is also the rake angle of the drill.

Shank—That part of the drill by which it is held and driven.

Straight Shank—A shank having the form of a cylinder. Usually pro vided on drills that are to be held in a chuck.

Taper Shank—A shank having an American Standard (Morse) Self Holding Taper.

Tang—The flattened end of a taper shank, intended to fit into a driving slot in a socket or a machine spindle.

Tang Drive—Two opposite, parallel driving flats on the extreme end of a straight shank.

Table 4-3. Recommended Cutting Speeds in Feet per Minute for Drilling
and Reaming Plain Carbon and Alloy Steels. See the Drilling and Reaming Columns.

Material AISI and SAE Steels	Hardness HB[a]	Material Condition	Cutting Speed, fpm HSS			
			Turning	Milling	Drilling	Reaming
Free Machining Plain Carbon Steels (Resulfurized)						
1212, 1213, 1215	100-150	HR, A	150	140	120	80
	150-200	CD	160	130	125	80
1108, 1109, 1115, 1117, 1118, 1120, 1126, 1211	100-150	HR, A	130	130	110	75
	150-200	CD	120	115	120	80
1132, 1137, 1139, 1140, 1144, 1146, 1151	175-225	HR, A, N, CD	120	115	100	65
	275-325	Q and T	75	70	70	45
	325-375	Q and T	50	45	45	30
	375-425	Q and T	40	35	35	20
Free Machining Plain Carbon Steels (Leaded)						
11L17, 11L18, 12L13, 12L14	100-150	HR, A, N, CD	140	140	130	85
	150-200	HR, A, N, CD	145	130	120	80
	200-250	N, CD	110	110	90	60
Plain Carbon Steels						
1006, 1008, 1009, 1010, 1012, 1015, 1016, 1017, 1018, 1019, 1020, 1021, 1022, 1023, 1024, 1025, 1026, 1513, 1514	100-125	HR, A, N, CD	120	110	100	65
	125-175	HR, A, N, CD	110	110	90	60
	175-225	HR, N, CD	90	90	70	45
	225-275	CD	70	65	60	40
1027, 1030, 1033, 1035, 1036, 1037, 1038, 1039, 1040, 1041, 1042, 1043, 1045, 1046, 1048, 1049, 1050, 1052, 1152, 1524, 1526, 1527, 1541	125-175	HR, A, N, CD	100	100	90	60
	175-225	HR, A, N, CD	85	85	75	50
	225-275	N, CD, Q and T	70	70	60	40
	275-325	Q and T	60	55	50	30
	325-375	Q and T	40	35	35	20
	375-425	Q and T	30	25	25	15
1055, 1060, 1064, 1065, 1070, 1074, 1078, 1080, 1084, 1086, 1090, 1095, 1548, 1551, 1552, 1561, 1566	125-175	HR, A, N, CD	100	90	85	55
	175-225	HR, A, N, CD	80	75	70	45
	225-275	N, CD, Q and T	65	60	50	30
	275-325	Q and T	50	45	40	25
	325-375	Q and T	35	30	30	20
	375-425	Q and T	30	15	15	10
Free Machining Alloy Steels (Resulfurized)						
4140, 4150	175-200	HR, A, N, CD	110	100	90	60
	200-250	HR, N, CD	90	90	80	50
	250-300	Q and T	65	60	55	30
	300-375	Q and T	50	45	40	25
	375-425	Q and T	40	35	30	15
Free Machining Alloy Steels (Leaded)						
41L30, 41L40, 41L47, 41L50, 43L47, 51L32, 52L100, 86L20, 86L40	150-200	HR, A, N, CD	120	115	100	65
	200-250	HR, N, CD	100	95	90	60
	250-300	Q and T	75	70	65	40
	300-375	Q and T	55	50	45	30
	375-425	Q and T	50	40	30	15

(continued)

Table 4-3. Recommended Cutting Speeds in Feet per Minute for Drilling and Reaming Plain Carbon and Alloy Steels. See the Drilling and Reaming Columns. (continued)

Material AISI and SAE Steels	Hardness HB[a]	Material Condition	Cutting Speed, fpm HSS			
			Turning	Milling	Drilling	Reaming
Alloy Steels						
4012, 4023, 4024, 4028, 4118, 4320, 4419, 4422, 4427, 4615, 4620, 4621, 4626, 4718, 4720, 4815, 4817, 4820, 5015, 5117, 5120, 6118, 8115, 8615, 8617, 8620, 8622, 8625, 8627, 8720, 8822, 94B17	125-175	HR, A, N, CD	100	100	85	55
	175-225	HR, A, N, CD	90	90	70	45
	225-275	CD, N, Q and T	70	60	55	35
	275-325	Q and T	60	50	50	30
	325-375	Q and T	50	40	35	25
	375-425	Q and T	35	25	25	15
1330, 1335, 1340, 1345, 4032, 4037, 4042, 4047, 4130, 4135, 4137, 4140, 4142, 4145, 4147, 4150, 4161, 4337, 4340, 50B44, 50B46, 50B50, 50B60, 5130, 5132, 5140, 5145, 5147, 5150, 5160, 51B60, 6150, 81B45, 8630, 8635, 8637, 8640, 8642, 8645, 8650, 8655, 8660, 8740, 9254, 9255, 9260, 9262, 94B30	175-225	HR, A, N, CD	85	75	75	50
	225-275	N, CD, Q and T	70	60	60	40
	275-325	N, Q and T	60	50	45	30
	325-375	N, Q and T	40	35	30	15
	375-425	Q and T	30	20	20	15
E51100, E52100	175-225	HR, A, CD	70	65	60	40
	225-275	N, CD, Q and T	65	60	50	30
	275-325	N, Q and T	50	40	35	25
	325-375	N, Q and T	30	30	30	20
	375-425	Q and T	20	20	20	10
Ultra High Strength Steels (Not AISI)						
AMS 6421 (98B37 Mod.), AMS 6422 (98BV40), AMS 6424, AMS 6427, AMS 6428, AMS 6430, AMS 6432, AMS 6433, AMS 6434, AMS 6436, AMS 6442, 300M, D6ac	220-300	A	65	60	50	30
	300-350	N	50	45	35	20
	350-400	N	35	20	20	10
	43-48 HRC	Q and T	25	…	…	…
	48-52 HRC	Q and T	10	…	…	…
Maraging Steels (Not AISI)						
18% Ni Grade 200, 18% Ni Grade 250, 18% Ni Grade 300, 18% Ni Grade 350	250-325	A	60	50	50	30
	50-52 HRC	Maraged	10	…	…	…
Nitriding Steels (Not AISI)						
Nitralloy 125, Nitralloy 135, Nitralloy 135 Mod., Nitralloy 225, Nitralloy 230, Nitralloy N, Nitralloy EZ, Nitrex I	200-250	A	70	60	60	40
	300-350	N, Q and T	30	25	35	20

[a] Abbreviations designate: HR, hot rolled; CD, cold drawn; A, annealed; N, normalized; Q and T, quenched and tempered; and HB, Brinell hardness number.

Speeds for turning based on a feed rate of 0.012 inch per revolution and a depth of cut of 0.125 inch.

Table 4-4. Recommended Cutting Speeds in Feet per Minute for Drilling and Reaming Tool Steels. See the Drilling and Reaming Columns.

Material Tool Steels (AISI Types)	Hardness HB[a]	Material Condition	Cutting Speed, fpm HSS			
			Turning	Milling	Drilling	Reaming
Water Hardening W1, W2, W5	150-200	A	100	85	85	55
Shock Resisting S1, S2, S5, S6, S7	175-225	A	70	55	50	35
Cold Work, Oil Hardening O1, O2, O6, O7	175-225	A	70	50	45	30
Cold Work, High Carbon High Chromium D2, D3, D4, D5, D7	200-250	A	45	40	30	20
Cold Work, Air Hardening A2, A3, A8, A9, A10	200-250	A	70	50	50	35
A4, A6	200-250	A	55	45	45	30
A7	225-275	A	45	40	30	20
Hot Work, Chromium Type H10, H11, H12, H13, H14, H19	150-200	A	80	60	60	40
	200-250	A	65	50	50	30
	325-375	Q and T	50	30	30	20
	48-50 HRC	Q and T	20	…	…	…
	50-52 HRC	Q and T	10	…	…	…
	52-54 HRC	Q and T	…	…	…	…
	54-56 HRC	Q and T	…	…	…	…
Hot Work, Tungsten Type H21, H22, H23, H24, H25, H26	150-200	A	60	55	55	35
	200-250	A	50	45	40	25
Hot Work, Molybdenum Type H41, H42, H43	150-200	A	55	55	45	30
	200-250	A	45	45	35	20
Special Purpose, Low Alloy L2, L3, L6	150-200	A	75	65	60	40
Mold P2, P3, P4, P5, P6	100-150	A	90	75	75	50
P20, P21	150-200	A	80	60	60	40
High Speed Steel M1, M2, M6, M10, T1, T2, T6	200-250	A	65	50	45	30
M3-1, M4, M7, M30, M33, M34, M36, M41, M42, M43, M44, M46, M47, T5, T8	225-275	A	55	40	35	20
T15, M3-2	225-275	A	45	30	25	15

[a] Abbreviations designate: A, annealed; Q and T, quenched and tempered; HB, Brinell hardness number; and HRC, Rockwell C scale hardness number.

Speeds for turning based on a feed rate of 0.012 inch per revolution and a depth of cut of 0.125 inch.

Table 4-5. Recommended Cutting Speeds in Feet per Minute for Drilling and Reaming Stainless Steels. See the Drilling and Reaming Columns.

Material	Hardness HB[a]	Material Condition	Cutting Speed, fpm HSS			
			Turning	Milling	Drilling	Reaming
Free Machining Stainless Steels (Ferritic)						
430F, 430F Se	135-185	A	110	95	90	60
(Austenitic), 203EZ, 303, 303Se, 303MA, 303Pb, 303Cu, 303 Plus X	135-185	A	100	90	85	55
	225-275	CD	80	75	70	45
(Martensitic), 416, 416Se, 416Plus X, 420F, 420FSe, 440F, 440FSe	135-185	A	110	95	90	60
	185-240	A,CD	100	80	70	45
	275-325	Q and T	60	50	40	25
	375-425	Q and T	30	20	20	10
Stainless Steels						
(Ferritic), 405, 409, 429, 430, 434, 436, 442, 446, 502	135-185	A	90	75	65	45
(Austenitic), 201, 202, 301, 302, 304, 304L, 305, 308, 321, 347, 348	135-185	A	75	60	55	35
	225-275	CD	65	50	50	30
(Austenitic), 302B, 309, 309S, 310, 310S, 314, 316, 316L, 317, 330	135-185	A	70	50	50	30
(Martensitic), 403, 410, 420, 501	135-175	A	95	75	75	50
	175-225	A	85	65	65	45
	275-325	Q and T	55	40	40	25
	375-425	Q and T	35	25	25	15
(Martensitic), 414, 431, Greek Ascoloy	225-275	A	60	55	50	30
	275-325	Q and T	50	45	40	25
	375-425	Q and T	30	25	25	15
(Martensitic), 440A, 440B, 440C	225-275	A	55	50	45	30
	275-325	Q and T	45	40	40	25
	375-425	Q and T	30	20	20	10
(Precipitation Hardening), 15-5PH, 17-4PH, 17-7PH, AF-71, 17-14CuMo, AFC-77, AM-350, AM-355, AM-362, Custom 455, HNM, PH13-8, PH14-8Mo, PH15-7Mo, Stainless W	150-200	A	60	60	50	30
	275-325	H	50	50	45	25
	325-375	H	40	40	35	20
	375-450	H	25	25	20	10

[a] Abbreviations designate: A, annealed; CD, cold drawn: N, normalized; H, precipitation hardened; Q and T, quenched and tempered; and HB, Brinell hardness number.

Speeds for turning based on a feed rate of 0.012 inch per revolution and a depth of cut of 0.125 inch.

Table 4-6. Recommended Cutting Speeds in Feet per Minute for Drilling
and Reaming Ferrous Cast Metals. See the Drilling and Reaming Columns.

Material	Hard-ness HB[a]	Material Condition	Cutting Speed, fpm HSS			
			Turning	Milling	Drilling	Reaming
Gray Cast Iron						
ASTM Class 20	120-150	A	120	100	100	65
ASTM Class 25	160-200	AC	90	80	90	60
ASTM Class 30, 35, and 40	190-220	AC	80	70	80	55
ASTM Class 45 and 50	220-260	AC	60	50	60	40
ASTM Class 55 and 60	250-320	AC, HT	35	30	30	20
ASTM Type 1, 1b, 5 (Ni Resist)	100-215	AC	70	50	50	30
ASTM Type 2, 3, 6 (Ni Resist)	120 175	AC	65	40	40	25
ASTM Type 2b, 4 (Ni Resist)	150-250	AC	50	30	30	20
Malleable Iron						
(Ferritic), 32510, 35018	110-160	MHT	130	110	110	75
(Pearlitic), 40010, 43010, 45006, 45008, 48005, 50005	160-200	MHT	95	80	80	55
	200-240	MHT	75	65	70	45
(Martensitic), 53004, 60003, 60004	200-255	MHT	70	55	55	35
(Martensitic), 70002, 70003	220-260	MHT	60	50	50	30
(Martensitic), 80002	240-280	MHT	50	45	45	30
(Martensitic), 90001	250-320	MHT	30	25	25	15
Nodular (Ductile) Iron						
(Ferritic), 60-40-18, 65-45-12	140-190	A	100	75	100	65
(Ferritic-Pearlitic), 80-55-06	190-225	AC	80	60	70	45
	225-260	AC	65	50	50	30
(Pearlitic-Martensitic), 100-70-03	240-300	HT	45	40	40	25
(Martensitic), 120-90-02	270-330	HT	30	25	25	15
	330-400	HT	15	–	10	5
Cast Steels						
(Low Carbon), 1010, 1020	100-150	AC, A, N	110	100	100	65
(Medium Carbon), 1030, 1040, 1050	125-175	AC, A, N	100	95	90	60
	175-225	AC, A, N	90	80	70	45
	225-300	AC, HT	70	60	55	35
(Low Carbon Alloy), 1320, 2315, 2320, 4110, 4120, 4320, 8020, 8620	150-200	AC, A, N	90	85	75	50
	200-250	AC, A, N	80	75	65	40
	250-300	AC, HT	60	50	50	30
(Medium Carbon Alloy), 1330, 1340, 2325, 2330, 4125, 4130, 4140, 4330, 4340, 8030, 80B30, 8040, 8430, 8440, 8630, 8640, 9525, 9530, 9535	175-225	AC, A, N	80	70	70	45
	225-250	AC, A, N	70	65	60	35
	250-300	AC, HT	55	50	45	30
	300-350	AC, HT	45	30	30	20
	350-400	HT	30	…	20	10

[a] Abbreviations designate: A, annealed; AC, as cast; N, normalized; HT, heat treated; MHT, malleabilizing heat treatment; and HB, Brinell hardness number.

Speeds for turning based on a feed rate of 0.012 inch per revolution and a depth of cut of 0.125 inch.

Table 4-7. Recommended Cutting Speeds in Feet per Minute for Drilling
and Reaming Light Metals and Copper Alloys

Material	Material Condition*	Cutting Speed, fpm		
		Drilling	Reaming	
Light Metals and Copper Alloys		HSS	HSS	Carbide
All Wrought Aluminum Alloys	CD	400	300	800
	ST and A	350	275	750
All Aluminum Sand and Permanent Mold Casting Alloys*	AC	500	350	900
	ST and A	350	275	750
*Except Alloys 390.0 and 392.0	AC	1 25	100	250
	ST and A	45	40	200
All Wrought Magnesium Alloys	A, CD, ST and A	500	350	100
All Cast Magnesium Alloys	A, AC. ST and A	450	375	1000
All Soft Brasses and Bronzes	A	160	160	320
	CD	175	175	360
All Medium Brasses and Bronzes	A	120	110	250
	CD	140	120	275
All Hard Brasses and Bronzes	A	60	50	80
	CD	65	60	200

*Abbreviations designate: A, annealed; CD, cold drawn; AC, as cast; ST and A, solution treated and aged.

Figuring Drill Turning Speeds

Many drill presses have charts attached to them that specify turning speeds for various materials in a range of drill diameters. As an alternative, use the information in Tables 4-3 through 4-7 and the formula below to determine the drill turning speeds.

Here's the formula:

$$N = \frac{12V}{\pi D}$$

where: N = Spindle speed of the drill press, rpm
(This is the speed at which the twist drill will operate.)
V = Cutting speed in feet per minute (fpm), at which the drill can cut the work material. Get this information from the Table.
The 12 represents 12 inches per foot.
π = pi = 3.1416. The mathematical symbol for this number which is constant for calculating the circumference of a circle.
D = Diameter of the drill, in inches. To convert fractions to decimals, see Appendix A.

Example 4-1:

Two holes are to be drilled in a part made from AISI 4140 steel that has been quenched and tempered to have a hardness of 270 HB (Brinell hardness number). The diameter of one hole is 1/8 inch and the diameter of the second hole is 1 inch. In Table 4-3 the cutting speed for this material is shown to be 60 fpm when the drill is made from high-speed steel.

Calculate the cutting speed for both drills.

For the 1/8 inch drill:

$$N = \frac{12V}{\pi D} = \frac{12 \times 60}{\pi \times \frac{1}{8}} = 1833 \text{ rpm}$$

For the 1 inch drill:

$$N = \frac{12 \times 60}{\pi \times 1} = 229 \text{ rpm}$$

The cutting speed in the metric system is given in terms of meters per minute (m/min) and the drill diameter is in millimeters (mm). Since the units are different, the cutting speed formula using SI units is different from the formula (4-1) in which customary inch units are used.

$$N = \frac{1000V}{\pi D}$$

(continued)

Where:

N = Revolutions per minute (rpm) at which the drill must operate.
V = Cutting speed, m/min
π = pi= 3.1416
D = Drill diameter, mm

Operating a Drill Press

In many cases, setting up means aligning the part on the machine and then clamping it in place or holding it securely by other means. The workpiece may or may not be clamped to the drill press table, depending on the size and shape of the workpiece and the size of the largest drill to be used. Small- and medium-size workpieces may be held against the table by hand—but not in the hand—or they may be clamped to the table. See Figure 4-4. When making this decision, it is always best to err in the direction of safety; i.e., when in doubt, secure the workpiece by holding it in a vise or by clamping it to the table.

Large-size drills (approximately 5/8 in. or 15 mm and larger) exert a very high torque which can only be resisted safely by clamping the workpiece to the table. The most common method of clamping is with U-shaped strap clamps.

When clamping a workpiece, the first step is to determine where the strap clamps should be placed. They must always be positioned so that the surface or surfaces to be machined can be reached by the cutting tools without interference from the clamps. Suitable surfaces on the workpiece are then selected on the basis of the ability to hold the part securely and the accessibility of T-slots or other openings in the table in which the clamping bolts—called T-bolts—can be anchored. This may also determine where on the machine tool table the workpiece is to be positioned.

Figure 4-4
Here's an example where the work can be safely held in place without the use of clamps.

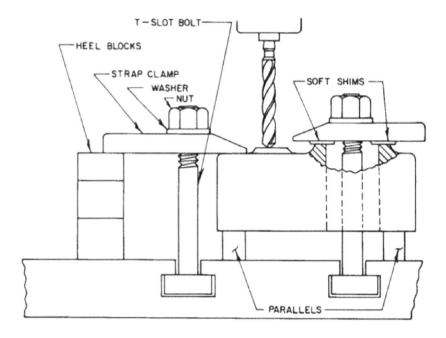

Figure 4-5 Typical drilling machine setup using strap clamps and T-bolts.

A typical setup for drilling is shown in Figure 4-5. In this setup the workpiece is placed on parallels to provide clearance for the drill when it has passed through the hole. The drill press table must always be protected from damage by the drill. If the holes to be drilled are blind holes which do not pass through the workpiece, it would be better to clamp it directly to the table. The T-bolt is placed in a T-slot or other suitable opening in the drill press table and the strap clamp is placed over the bolt. The open end of the strap clamp is placed on the workpiece and the other end on heel blocks of suitable height so that the clamp is parallel with the table. A washer and a nut are then placed on the bolt and tightened. In some instances, the T-bolt may be placed through an opening in the workpiece, as shown by the right-hand clamping arrangement in Figure 4-5. In this case heel blocks may not be required. Finish machined surfaces on the workpiece must be protected from damage by placing soft metal shims between the strap clamp and the workpiece.

Drill Press Vises. When using small drills (approximately 5/16 in. or 8 mm and smaller), the workpiece does not usually need to be clamped, although it may have to be held in a drill press vise to provide the leverage required to safely overcome the drill torque. Drill press vises are a necessary accessory in drilling. They are used to hold workpieces when drilling large, medium, and small holes. When drilling with large and medium-size drills, the vise may be clamped to the table while holding the workpiece. Some drill press vises have a V-groove machined on the solid jaw, which is used to locate and hold round workpieces, as shown in Figure 4-7.

Correct and Incorrect Use of Strap Clamps

Several correct and incorrect applications of strap clamps are illustrated in Figure 4-6. As shown in view A, the T-bolt should be placed as close to the workpiece as possible. It should always be placed closer to the workpiece than to the heel blocks to assure that the greater proportion of the clamping force applied by the bolt and nut is applied to the workpiece and not to the heel block. Also shown in this view is the principle that the T-bolt should not be longer than necessary. As already mentioned and shown in view B, the height of the heel blocks should be such that the strap clamps are parallel with the machine tool table. Parallels, when used, should be placed directly below the strap clamps, as shown in view C, to prevent the clamping force from causing the workpiece to bend or spring out of shape. Otherwise, when the clamping force is released, the workpiece will spring back to its original shape causing the surfaces that were machined while the part was clamped to be inaccurate. Any overhang of the workpiece below a clamp must he supported for this reason. When the opening of the overhang is large, a screw jack may he used as a support against springing the part, as shown in view D. When the opening is narrow, shims and wedges are used, as shown in view E.

Figure 4-6
Correct and incorrect applications of strap clamps.

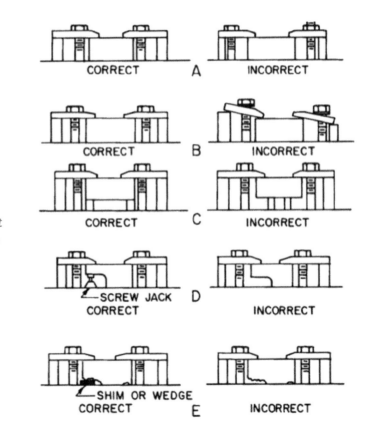

Figure 4-7 A drill press vise holds a workpiece during drilling.

Drilling the Hole

Except when using high production and computer controlled machines, drills less than approximately ¼ inch, or 6.4 mm, are usually fed through the workpiece using the manual controls on the drilling machine, because the operator can quickly develop a sense of feel and sight that will provide the correct feed rate and prevent drill breakage. See Figure 4-8. Even drills, up to one-half inch, or 13 mm, are often preferably fed "by hand." The feed rate is dependent on the workpiece material (type, hardness, and heat treatment), the size of the drill, and the rigidity of the machine, the workpiece, and the setup. All of these factors must be considered in selecting the feed rate. As a guide, Table 4-8 provides a range of feed rates recommended for different size twist drills.

Drill Press Safety

Always wear safety glasses when operating a drill press, and remember to keep long hair and loose fitting clothing away from the work area. Be sure to unplug the press when changing drill bits or changing speeds.

The conditions influencing the cutting speed and feed may vary in different shops. Moreover, all of the factors may not be known; e.g., the hardness and heat treatment of the work material may not be known exactly. In such cases judgment should be used in applying the values in the tables. These values do, however, provide a good starting point from which changes can be made as experience is gained.

Table 4-8. Recommended Feeds in Inches per Revolution for High-Speed Steel Twist Drills

Drill Diameter, inch	Feed, in./rev	Drill Diamete, inch	Feed, in./rev
1/16 to 1/8	.001-.003	Over 1/2 to 3/4	.005-.012
Over 1/8 to 1/4	.002-.006	Over 3/4 to 1	.006-.015
Over 1/4 to 1/2	.004-.010	Over 1	.010-.025

Figure 4-8 Drill presses used in the home shop are usually operated manually.

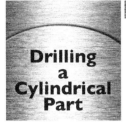

Drilling a Cylindrical Part

A job that is frequently encountered is that of drilling a hole through the side of a cylindrical part, as shown in Fig. 4-9. The scribed layout line on the end face of the workpiece is aligned in a vertical position-by the blade or the beam of a combination square (A and B, Fig. 4-9). The work is then clamped in position. The square is not used to align the drill with the work piece. Very large cylindrical parts can sometimes be aligned in the manner shown at D. The blade of a combination set has a center head and a square head fastened to it, as shown. The blade is positioned over the punch mark at the center of the hole. Sometimes, greater accuracy can be obtained by placing the blade opposite the layout line as shown at D. The workpiece is adjusted until the spirit level on the square head is exactly level. The part is then clamped into position.

Figure 4-9
(Upper left) *Using the blade of a combination square to align cylindrical part.*
(Upper right)*Using the beam of a combination square to align a cylindrical part*
(Lower)*. Using a center head combined with a square head to align a large cylindrical part.*

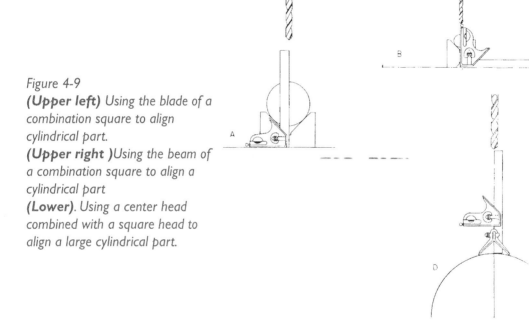

Aligning the Drill

The first step is to align the drill with respect to the center punch mark at the center of the hole to be drilled. The drill point is then fed into the workpiece until it has penetrated approximately two-thirds of the drill's diameter. The drill is then backed away and the relation of the drilled spot to the scribed layout circle should be observed.

Fixing an Off-Center Hole

As shown in Figure 4-10 A, the spot produced by the drill point is not perfectly centered in the layout circle, indicating that if the hole were to continue to be drilled in this position it would not be in the desired location. To correct this situation, cut one or more shallow grooves on the heavy side, as shown at B of Figure 4-10, with a gouge, or draw chisel (Figure 4-11). The groove unbalances the cutting forces acting against the lip of the drill causing it to shift in the direction of least resistance, which is toward the grooved side. The drill point is fed a little deeper into the work, and the location between the spot and the layout circle is again checked. If necessary, additional grooves are cut with a draw chisel. When the spot and the layout circle are concentric, as shown at C of Figure 4-10, the hole can be drilled to the required depth. When the hole has been drilled, about one-half of each witness mark will remain to show that the hole is drilled exactly where the layout has indicated that it be made. It should be mentioned that it is usually not necessary to cut grooves in every hole as described. Frequently, the initial check between the spot and the layout circle will show that they are concentric and the hole can be drilled to depth without additional positioning of the work.

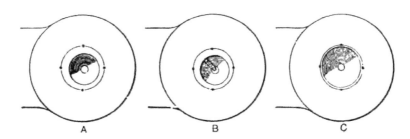

Figure 4-10 Method of starting drill, concentric with scribed circle.

Center Drill. An alternate method of starting a drilled hole is to center drill a hole first (see Figure 4-12). The center-drilled hole is then used as a guide for starting the twist drill. The center drill is short and stiff, and will not bend. Furthermore, it has a relatively small chisel edge when compared to larger size drills, thereby enabling the center drill to penetrate the workpiece with greater ease. The coned surface formed by the center drill has an included angle of 60° which is steeper than the included angle of the drill point. Thus, the lips of the twist drill contact the workpiece first and the chisel edge does no work until the location of the hole is established. This method is especially recommended for starting holes on machines where the work is normally clamped to the table and where the spindle can be aligned exactly to the desired location on the workpiece. When this method is used on an upright or sensitive drilling machine, the workpiece should not be clamped when starting the hole with the center drill. Allowing the workpiece or spindle to "float" lets the center drill find and align itself with the center punch mark as it starts drilling into the workpiece. After the workpiece and the machine spindle are aligned, and when the center hole has been drilled, the workpiece may be clamped, when necessary, to drill the hole to size.

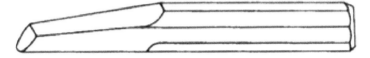

Figure 4-11 Draw chisel or gouge.

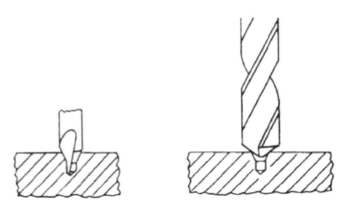

Figure 4-12 Starting a drilled hole with a center drill.

Hole Location Transfer

In machine shop work the location of a hole must frequently be transferred from one part to another. One way to accomplish this is with the use of a transfer punch.

Transfer punches are cylindrical in shape to fit snugly into the hole in which they are to be used. One end is hardened and has a small sharp punch point that is, ground concentric with the body. The other end must be soft—this is very important in order to prevent possible serious injury when this end is struck with a hammer. Transfer punches can be purchased commercially, or they can be made on the job from drill rod or other types of tool steel. The use of transfer punches is illustrated in Figure 4-13. After the two parts are located and clamped together; the transfer punch is inserted in the hole and tapped lightly with a hammer. The parts are then disassembled, and the transfer punch mark is very carefully enlarged slightly with a regular center punch. A start hole is then drilled into the workpiece using a 1/32 inch or a 0.75 mm drill. Successively larger holes are then drilled into the workpiece to enlarge it to the required size and depth. This procedure will give the most precise results. If less accuracy is satisfactory, a circle can be scribed with dividers, witness marks punched on the circle, and the hole drilled in the usual manner.

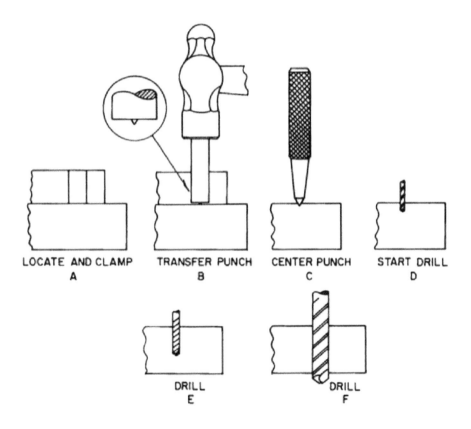

Figure 4-13 Procedure for using a transfer punch.

Reamers

Reamers are used to finish cut holes to a precise dimension and to provide a smooth surface finish on the walls of the hole. They are precision cutting tools that must always be handled with care to prevent chipping and other damage to the reamer teeth. *One cardinal principle in using a reamer is never to rotate it in the reverse direction.* Reamers should always be rotated in the for-

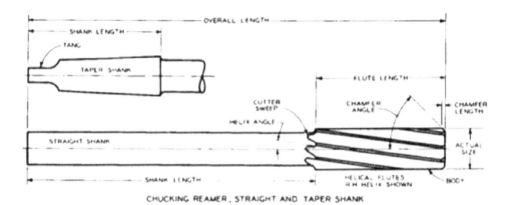

CHUCKING REAMER, STRAIGHT AND TAPER SHANK

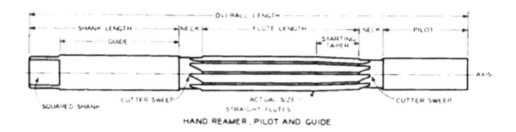

HAND REAMER, PILOT AND GUIDE

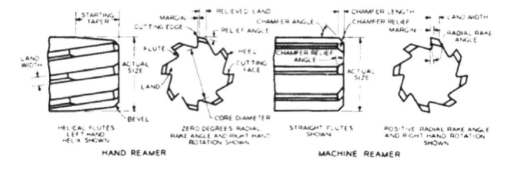

HAND REAMER   MACHINE REAMER

Figure 4-14 Reamer nomenclature.
Courtesy of the Metal Cutting Tool Institute

ward, or cutting direction, at all times; regardless of whether they are entering or leaving the hole. Failure to do this will rapidly damage the margin and destroy the accuracy of the reamer. Reaming is a fast, efficient, and easily applied method of finishing holes.

There are two basic types of reamers: machine reamers and hand reamers. See Figure 4-14. Machine reamers may have a straight, or a taper, shank. They are used primarily in power-driven machine tools. Machine reamers have a chamfer ground on the ends of the lands. The chamfer angle is usually 45°, although other chamfer angles are sometimes used. The lands have a margin adjacent to the cutting edge, which is unrelieved. Relief is ground on the back of the chamfer, thereby making it into a sharp cutting edge. The chamfer edges do the cutting and remove the excess metal from the hole. The forces on the workpiece resulting from the cutting action of the reamer will cause the work to spring slightly away from the reamer, in the region of the chamfer edge. After the chamfer edge has passed a given portion of the hole, the work material will spring back and bear against the margin of the lands of the reamer. The margin, being unrelieved, is a poor cutting edge. Primarily, it burnishes the sides of the hole to produce a good surface finish. It may also scrape a very small amount of metal out of the hole, thereby removing small irregularities. The margin also performs another important function in supporting and steadying the reamer as it passes through the hole. Sometimes, the margin is ground with a slight taper toward the chamfer edge. This is sometimes called a lead. Actually, its function is to increase the burnishing action of the lands, thereby improving the surface finish of the hole.

Hand reamers are designed to be turned manually in reaming a hole. They are ground with a starting taper on the lands which is relieved to form a sharp cutting edge. (See Figure 4-14.) A bevel edge is also ground on the end, however most of the cutting is done by the starting taper.

Hand reamers are designed to remove less metal from a hole than machine reamers will. The margin of the hand reamer also burnishes the hole in the same manner as does the margin on the machine reamer. The speed and feed through the hole of a hand reamer are controlled entirely by the workman. On machine reamers, the speed and feed are controlled by the machine setting.

Cutting Speed. As a general rule, the cutting speed for reaming should be about one-half to two-thirds of the speed used for drilling the same material. Excessive cutting speeds will cause the reamer to chatter; consequently, the cutting speed must be kept low enough to prevent the occurrence of chatter. The feed for machine reaming is usually .0015 to .0040 inch per flute per revolution. Thus, the feed rate of a six-fluted reamer that is cutting at 100 rpm, using .004 inch per tooth feed, would be 6 × .004, or .024 inch per revolution. Expressed in different terms, the feed rate would be 100 × .024, or 2.4 inches per minute. When the feed used is too low the work may be glazed, the cutting edge will wear excessively, and, occasionally, chatter will occur. An excessively fast feed rate will reduce the accuracy of the hole and lower the quality of the surface finish.

Using Reamers

Reamers are used to finish drilled holes to size and to a smooth, surface finish. In order to produce accurate holes by reaming, the reamer must be sharp and correctly ground, the right amount of stock must be left in the hole for reaming, and the reamer must be accurately aligned with the axis of the drilled hole. The feed of the reamer through the hole must not be too fast (.0015 to .004 inch per land), otherwise the accuracy and the surface finish of the hole will be impaired. Sometimes a reamer will chatter. This chatter may exist only at the start of the hole as the reamer aligns itself with the drilled hole. Chatter is objectionable because it impairs the surface finish of the hole and it can damage the reamer. Cemented-carbide tipped reamers, in particular, are quite likely to chip if chatter occurs even at the start of the hole.

Correcting Chatter

To eliminate chatter, try these steps:

1. Reduce the cutting speed.
2. Vary the feed.
3. Reduce the relief angle on the reamer.
4. Use a piloted reamer.
5. Use a reamer having helical flutes.
6. Iincrease the rigidity of the setup.
7. Use a reamer having irregular spacing of the flutes.

The use of a helical fluted reamer is much preferred to one having irregular spacing. The helix angle of helical fluted reamers should not be greater than necessary, as this will tend to increase the force required to feed the reamer through the hole. One exception to this are taper reamers; these will cut better with a very steep spiral. Often, the difficulties associated with chatter can be eliminated by exercising great care in aligning the reamer with the hole at the start of the operation. This should never be done carelessly.

Cutting Fluid. A good supply of the correct cutting fluid should be used for the reaming operation except when reaming cast iron, which should be reamed dry, with only a jet of compressed air to act as a cooling medium. Use of the correct cutting fluids often materially improves the surface finish obtained in a reamed hole. Generally, for reaming steel, a sulfurized mineral oil will work best although good results have also been obtained using soluble oil. Soluble oil and lard oil compounds can be used on aluminum. Brass may be reamed dry, although the application of a soluble oil or a lard base, oil compound is often advisable. Soluble oil, or a sulfurized mineral oil, is recommended for stainless steels. Many proprietary cutting fluids can be purchased which will do an excellent job.

Counterbores, Countersinks, and Spotfacers

Counterbores, countersinks, and spotfacers are frequently used in drill-press work in order to modify an existing hole. The counterbore, A, in Figure 4-15, is used to enlarge the end of a hole by cutting a cylindrical surface that is concentric with the original hole. The concentricity is obtained by the pilot located on the end of the counterbore and which acts to guide the tool as it penetrates into the work. A countersink, B, in Figure 4-15, is used to make a cone-shaped enlargement in the end of the hole. A spotfacing tool is shown at C. The purpose of the spotfacing operation is to cut a smooth, flat surface which is perpendicular to the axis of the hole. Usually, this surface serves as a seat for the head of a cap screw or bolt, or as a seat for a nut.

Figure. 4-15
A. Counterboring a hole.
B. Countersinking a hole.
C. Spot-facing a hole.

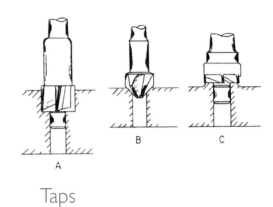

Taps

The basic nomenclature of a tap is shown in Figure 4-16. A tap is used to produce internal threads. It has thread-shaped teeth which cut a thread of similar form when it is screwed into a drilled hole. The first step in tapping an internal thread is to drill a hole with a tap drill. This is an ordinary drill that is appropriately sized so that it will leave just the right amount of metal in the hole for the tap to remove and finish the thread forms to required size. The tap is then screwed into the workpiece by hand, with the aid of a tap wrench, or by power, as applied by a machine tool such as a lathe or a drilling machine. The feed of the tap must always be equal to the lead of

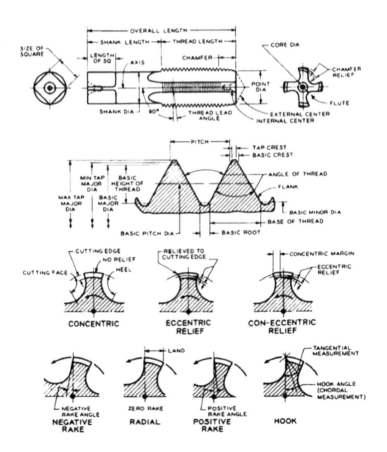

Figure 4-16
Basic nomenclature of a tap.
Courtesy of the Metal Cutting Tool Institute

the thread being cut. The lead of a thread is the distance that it advances in one revolution. A tap will generally feed itself, once it has started, as it will tend to follow that portion of the thread that already has been cut in starting.

Hand taps have a cylindrical shank and a driving square on the end to which a tap wrench can be attached. These taps are used for most machine tapping applications, as well as for hand tapping. They can be obtained as ground taps, which are finished by precision grinding, or as cut taps which are finished by a cutting operation.

Hand taps are classified into the following types: taper, plug, and bottoming. Each type is distinguished by the amount of chamfer on the end. The angle formed between the chamfer (see Figure 4-16) and the axis of the tap is called the chamfer angle. It determines the number of threads having their height reduced on the end of the tap. The chamfered threads are relieved behind the face and do all of the cutting to form the thread on the part. The number of flutes and the length of the chamfer determine the chip load, or the amount of material removed by each tooth. See Figure 4-18.

A cutting fluid should always be used with all materials when tapping except on cast iron and plastics. When hand tapping a hole, the tap should be advanced approximately one turn and then rotated in the reverse direction until the chip is broken. By repeating this sequence until the hole is tapped, a smooth thread will be produced with minimum effort. Machine-driven taps are generally screwed into the hole while rotating in the forward direction and are reversed to remove the tap when the job is finished. On very difficult jobs the tap is sometimes reversed when cutting into the hole in a manner similar to hand tapping.

Using Taps

When tapping is to be done in a drilling machine there must be provision for rapidly reversing the spindle so that the tap can be backed out of the hole after the thread is cut to the required depth. Some bench drill presses are not equipped with such a mechanism for reversing the rotation of the spindle, but attachments to do so are available. An automatic reverse tapping attachment is shown in Figure 4-17. This attachment has a taper shank which is inserted in the spindle of the

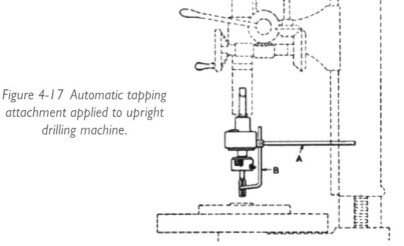

Figure 4-17 Automatic tapping attachment applied to upright drilling machine.

machine. The body of the chuck and the gage B, which may be used to control the depth to which holes arc tapped, are both prevented from rotating with the machine spindle by rod A. When tapping a hole, the spindle is lowered by hand and the tap is fed to the required depth manually. When the lower end of the stop rod B comes into contact with the face of the work, the forward rotation of the tap is stopped. When the drill press spindle is raised, gears within the chuck body are engaged that cause the tap to back out of the hole rapidly. The tapping attachment may be equipped to have an adjustable friction drive that stops the rotation of the tap whenever the tapping torque exceeds a preset limit, thus safeguarding the taps against breakage.

There are many more variables that must be considered when selecting the cutting speed for tapping than for other machining operations; thus, it is not surprising that the best and most efficient cutting speeds for tapping cannot be tabulated with any assurance that the tap will then be operating at its highest efficiency. Among the factors that must be considered are:

1. Material to be tapped
2. Heat treatment and hardness of material to be tapped
3. Length of hole
4. Length of chamfer on top
5. Pitch of thread
6. Percentage of full thread to be cut
7. Cutting fluid to be used
8. Design of the drilling machine with respect to the sensitivity of the controls

For example, speeds must be lowered as the length of the hole is increased because long holes accumulate chips, increase friction, and interfere with the flow of cutting fluids. Taps having long chamfers can be run faster than taps with short chamfers in short holes; however, in long holes the taps with the short chamfers can be run faster. Bottoming taps must be run slower than plug taps. A slower speed is required to tap full-depth thread than a 75 percent-depth thread. When the tap diameter exceeds approximately ½ inch, coarse thread taps should be run more slowly than fine thread taps having the same diameter. Taper threaded taps, such as pipe taps, should be operated from one-half to three-quarters of the speed of a straight tap of comparable diameter. Automatic machines in which the movements required for tapping are machine controlled, can, in most cases, be operated at greater tapping speeds than manually operated machines. The machine operator must exercise good judgment in deciding the limits of his ability to control the machine in order to prevent tap breakage and other possible damage. Table 4-9 provides recommended tapping speeds, but these may have to be modified by the factors discussed above.

Cutting Fluid. When tapping, a cutting fluid should always be used with all materials except cast iron and plastics. Cast iron and plastics are machine tapped dry or with a jet of compressed air to cool the tap and help to remove the accumulation of chips. It is not practical to give specific recommendations for tapping other materials because of the wide variety of proprietary cutting fluids that can do an excellent job. The following recommendations are general and it should be understood that there are exceptions. Sulfur based oils with active sulfur are recommended for tapping plain carbon steels, alloy steels, malleable iron, and Monel metal. Tool steels, high speed steels, stainless steels, and alloy steels that have been heat treated to a higher hardness should be tapped using achlorinated sulfur base oil. Mineral oil with a lard base is recommended for aluminum, brass, copper, manganese bronze, naval brass, phosphor bronze, and Tobin bronze. Aluminum and zinc die castings should be tapped with lard oil that has been diluted with 40 to 50 percent kerosene. The proper cutting fluid used when tapping can increase tap life, improve surface finish, reduce friction, and result in better size control.

Table 4-9 Cutting Speeds for Machine Tapping

Material	Cutting Speed, fpm	Material	Cutting Speed, fpm
Low Carbon Steels Up to .25% C	40 to 80	*Aluminum*	50 to 200
		Brass	50 to 200
Medium Carbon Steels .30 to .60% C		*Manganese Bronze*	30 to 60
Annealed	30 to 60	*Phosphor Bronze*	30 to 60
Heat Treated (220 to 280 Bhn)	20 to 50	*Naval Brass*	80 to 100
Tool Steels, High Carbon &		*Monel Metal*	20 to 40
High-Speed Steel	20 to 40	*Tobin Bronze*	80 to 100
Stainless Steels	5 to 35	*Plastics*	
Gray Cast-Iron	40 to 100	Thermoplastics Thermosetting	50 to 100 50 to 100
Malleable Iron Ferritic	80 to 120	*Hard Rubber*	50 to 100
Pearlitic	40 to 80	*Bakelite*	50 to 100
Zinc Die Castings	60 to 150		

Figure 4-18 Threads being produced by a manual tap.

Common Problems with Drilled Holes

Text and photos reprinted by permission of
Sandro Di Filippo
and *The Home Shop Machinist*
(November/December 2012).

When I was an apprentice toolmaker, I would get upset when my drilled holes weren't perfect. They would be out of round, oversize, or crooked. What had gone wrong? How had I screwed up something so simple? Years went by before it dawned on me that drilling a hole was something like using a hacksaw; they are meant to remove material quickly, but not with any real precision.

Since drilling is such a common practice in the machine shop, I thought it would be a good topic for this month's issue. I'm going to limit this discussion to twist drills, so-called because of their spiral flutes. Straight flute drills also exist and are used for materials that have a tendency to grab, such as copper or brass. Regular drills can be used for the same purpose with a slight modification, which I'll discuss later.

Drills are available in two styles: straight shank and taper shank. Drills smaller than 1/2″ are more convenient to use with a straight shank. Straight shank drills larger than 1/2″ do exist, but you'll need a larger drill chuck to hold them. A convenient solution is to use 1/2″ shank drills with a larger body size, sometimes called Silver & Deming drills. They are usually available up to 1″ in diameter and can be held in a normal 1/2″ drill chuck. They typically have three flats at 120° around the shank for the jaws to bear on to prevent the drills from spinning in the chuck. Caution: Never use Silver & Deming drills in a keyless chuck. The increased torque can tighten the chuck enough to split the outer ring; I learned that one the hard way!

Taper shank drills are usually a Morse taper and are mostly used for drills larger than 1/2″, but are available in smaller sizes. The tang is soft and is only for knocking out the drill; it is not meant to drive the drill, the taper does that. Sleeves are used to step up the size of a smaller taper so it will fit in a machine with a larger one. When inserting a drill into the spindle, be sure to firmly seat the drill. If it doesn't stay in, check the taper for any bruises or burrs, which can be cleaned up with a fine cut file. A quick look through *Machinery's Handbook* or a little time on the internet will turn up plenty of examples of the different drill styles.

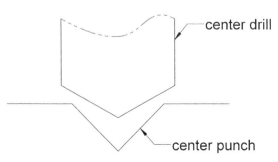

Figure 1

Some common problems with drilling a hole are: the drill wandering out of location, drilling a crooked hole, and drilling an oversize hole. None of these problems are that difficult to prevent, it just takes a bit of understanding and some patience to get things right.

The drill wandering out of location is probably the biggest problem, so let's start with that one. The web of a drill does not cut; in fact, the very center of the drill is pushing or displacing the metal ahead of it. If you were to start a drill in just a center punched mark, the web of the drill may not center itself properly, resulting in the drill starting the hole in the wrong location. After laying out and center punching a hole location, normal procedure is to start the hole with a center drill. A center drill is very short and stiff and, in theory, will locate the work under it by aligning itself in the cone of the punch mark. My own experience has taught me that this doesn't always happen.

The problem is the size of the punch mark compared to the size of the center drill (*Figure 1*). The center drill is normally larger than the punch mark and doesn't always find its way to the correct location. My solution is to start with a drill smaller than the punch mark (*Figure 2*) and create a dimple in the bottom of it (*Figure 3*). Next, open up the dimple with a slightly larger drill, about 1/8″, and repeat until it is bigger than the web of the final drill size (*Figure 4*). I've never had a problem with the drill wandering off center with this procedure.

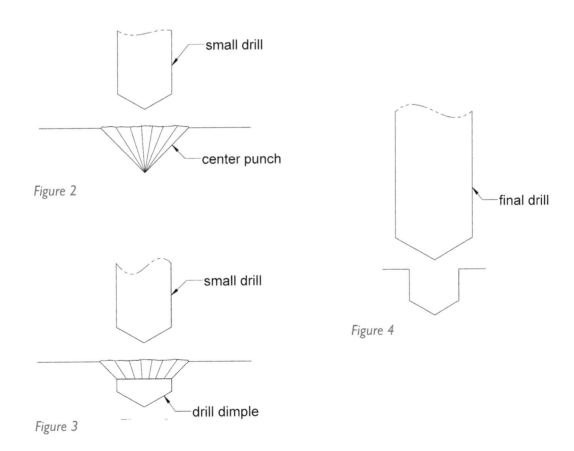

Figure 2

small drill

center punch

Figure 3

small drill

drill dimple

Figure 4

final drill

A crooked hole is a common problem with deep holes, and more difficult to prevent, since you don't know about it until you've finished drilling the hole. The trouble stems from the flexibility of twist drills and the non-cutting action of the web of the drill. (*Figure 5*) gives a much exaggerated representation of what happens. Because the web doesn't cut, applying too much pressure on the drill will cause the body to bend and push the point over to one side.

It's not a difficult problem to prevent, as long as you use the right procedure. If you need to drill a hole right through a long piece, the simplest solution is to drill from both sides. A blind hole isn't that big of a deal if you approach it correctly. One solution is to start off with a shorter drill first and drill as deep as you can before following with a longer drill. The shorter drill will be stiffer and this will help prevent a crooked hole. Another solution is to drill a pilot hole that is slightly bigger than the web of the larger drill, thus reducing the tendency of the drill to bend. In all cases, withdraw the drill frequently to clear chips, and run the drill at the correct speed.

Slightly oversize holes are normal when drilling. Very oversize holes are almost always caused by an incorrectly sharpened point. The only important thing to remember when sharpening a drill is to keep both lips at the same angle and the same length (*Figure 6*). But

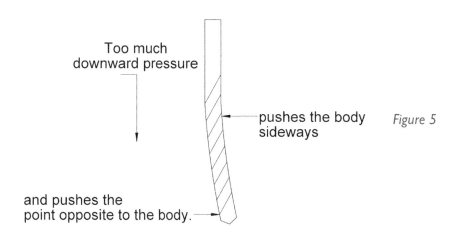

Too much downward pressure

pushes the body sideways

Figure 5

and pushes the point opposite to the body.

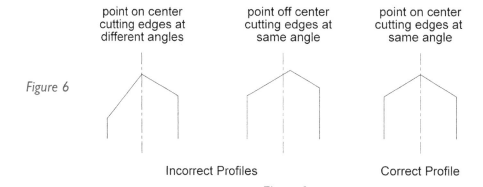

Figure 6

point on center cutting edges at different angles

point off center cutting edges at same angle

point on center cutting edges at same angle

Incorrect Profiles

Correct Profile

what if you need a hole to be very close to size and you don't have a reamer, can you drill it? Yes, with a slight modification to the drill. First drill the hole like I've described, but leave it undersize by about 1/32″ to 1/16″, depending on the size of the hole. Next, an ordinary drill is modified by rounding over the corners *(Photo 1)* and honing a small flat on the cutting edge *(Photo 2)*. The rounded corners help the drill to center itself in the hole, and the flat prevents the drill from grabbing.

Let me explain that last point. The flat is essential to prevent the drill from pulling itself into the hole. A drill looks and acts just like a wood screw. A wood screw bites into the wood and will draw itself into the wood. A drill can do the same. Normally there is enough pressure on the web of the drill to prevent it from grabbing the work and pulling itself in. When opening up a hole, there isn't enough pressure on the end to prevent it from grabbing. By putting a flat on the cutting edge, the cutting pressure is increased and prevents it from grabbing. This trick also works with "grabby" materials like copper or brass.

Once the drill is modified, proceed just like a reamer. Run the drill at about 1/4 of the normal cutting speed using some cutting oil. The results aren't quite as good as a reamer, but it will give you a nicely round and on-size hole without too much trouble. If you're not sure about cutting speeds, refer to my previous article, "Cutting Speed and Feed Rates" *(The Home Shop Machinist,* May/June 2012*)*.

Finally, let's talk about the setup to drill a hole. *(Photo 3)* shows a typical drilling setup with the piece held in a vise. Never attempt to hold the part with just your hand; always clamp the piece to the table or hold it in a vise. For added security, I have a thick washer bolted to the table for the vise to rest against while drilling. This helps to resist the twisting forces of the drill and keeps the work on the

Photo 1

Photo 2

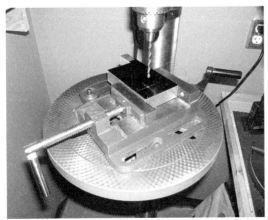

Photo 3

Photo 4

table of the drill press. I like to use a foot switch with my drill press *(Photo 4)*. By walking away (or running, if necessary) I can quickly turn it off, leaving my hands free to operate the machine.

As I said at the beginning, drilling a hole isn't complicated, but like any other machine operation it does have its quirks that only become apparent with experience. So don't be afraid to drill a few holes, just use some care and you should have no problems.

5 Introduction to Lathes

To many, the lathe is the heart of any machine shop. This relatively simple tool makes it possible to cut and shape all metals. Basically, the tool holds and turns the material while a stationary, though adjustable, cutting tool performs a variety of cutting and shaping functions, including turning cylindrical surfaces, facing flat surfaces, cutting threads, drilling and boring holes, among others.

Of equal importance in considering the versatility of the lathe is the ease with which the workpiece can be set up and held, and the ease with which the lathe performs different operations. When operated with skill and with an understanding of its capabilities, the lathe can produce parts to exacting standards of accuracy and finish. With the appropriate size machine, the lathe can produce both large and small parts.

Principal Parts

The principal parts of a bench-top lathe are shown on the following page in Figure 5-1. The bed provides a sliding surface for the *saddle* and *cross-slide*. The saddle and cross-slide hold the cutting tool in place. A stable bed keeps the *headstock* and *tailstock* aligned with one another. The bed on good-quality tools provides for the precise alignment of the headstock and the *tailstock*, which is important because the headstock contains the spindle that holds and turns the workpiece; the tailstock supports long material and is used for drilling.

The Motor

The lathe is driven by an electric motor that is usually located at the headstock end of the tool. A multiple V-belt drive transmits the power from the motor to the headstock. The principal function of the headstock is to support and to align the headstock spindle. On many lathes, it also contains a selective gear transmission that provides for different speeds of the lathe spindle. Although the specifications for different products vary, bench-top lathes tend to operate up to 2500 to 2800 rpm.

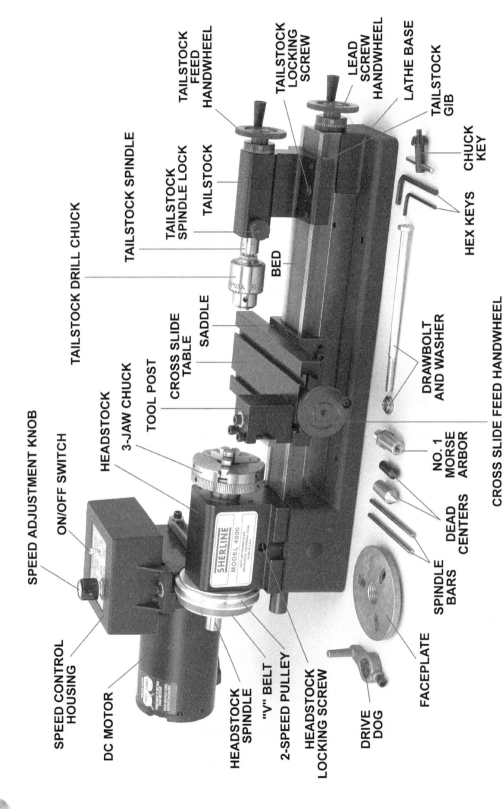

TAILSTOCK FEED HANDWHEEL

TAILSTOCK LOCKING SCREW

LEAD SCREW HANDWHEEL

LATHE BASE

TAILSTOCK GIB

CHUCK KEY

HEX KEYS

TAILSTOCK SPINDLE

TAILSTOCK SPINDLE LOCK

TAILSTOCK

TAILSTOCK DRILL CHUCK

BED

SADDLE

CROSS SLIDE TABLE

TOOL POST

3-JAW CHUCK

HEADSTOCK

DRAWBOLT AND WASHER

NO. 1 MORSE ARBOR

DEAD CENTERS

SPINDLE BARS

FACEPLATE

DRIVE DOG

HEADSTOCK LOCKING SCREW

"V" BELT

2-SPEED PULLEY

HEADSTOCK SPINDLE

DC MOTOR

SPEED CONTROL HOUSING

SPEED ADJUSTMENT KNOB

ON/OFF SWITCH

CROSS SLIDE FEED HANDWHEEL

SHERLINE
MODEL 4000

Figure 5-1 Here is a bench-top lathe with all of its principal parts labeled. Courtesy of Sherline Products

The Carriage

The carriage, or saddle, supports the cross slide and the tool holder. See Figure 5-2. The saddle is an H-shaped casting that slides on the top of the lathe bed. It enables the cutting tool to be moved parallel or perpendicular to the axis of the lathe. The operator can move the saddle toward or away from the headstock. The saddle rides on of the ways on the lathe bed, which guides the movement of the carriage parallel to the lathe axis.

The Cross Slide. The cross slide moves on top of the saddle from one side of the saddle across to the other side. The movement of the cross slide is perpendicular to the lathe axis. This movement is actuated by a feed screw. A dial shows the position of the cross slide. The feed screw can be turned manually by means of the cross-feed handle. Some bench-top lathes contain a compound rest. This is a tool holder that allows the operator to position the cutting tool at any angle, which is useful for cutting tapers. The manufacturers of models that do not come with a compound rest usually offer them as an accessory.

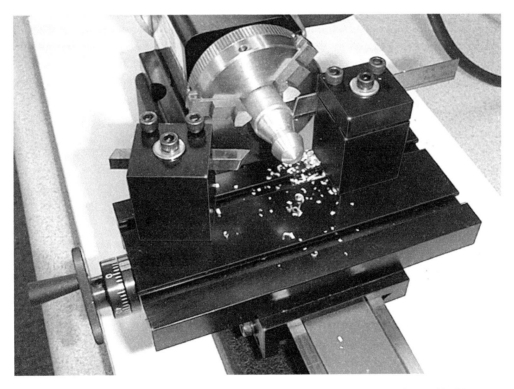

Figure 5-2 The saddle moves across the width of the lathe to place the tool holder in the proper position. Courtesy of Sherline Products

The Tailstock

The tailstock supports long workpieces and also holds drills, reamers, and other cutting tools. It consists of a base that rests on the bed of the lathe and a body that is clamped to the base.

The base, which is aligned with the axis of the lathe by one of the ways of the lathe, has an accurately machined guide that fits the body. The body and the base will slide over the lathe bed until it is in the desired position, then it is clamped to the bed by tightening the clamp bolt nut. The tailstock spindle is prevented from turning by a key. When the tailstock handwheel is turned, the screw causes the tailstock spindle to slide in or out of the body. The spindle may be clamped by turning the binder handle. The inside of the spindle nose has a taper that holds the tailstock center, drill chucks, or cutting tools such as drills or reamers. This taper must be absolutely clean before a center or other tool is inserted in it. It serves, primarily, to locate; its ability to resist torsion is very limited. It should not be used to hold large drills because the cutting torque of the drill may cause the taper to slip and possibly score the surface. When this occurs, the taper will no longer locate the center or the tool accurately, and it will no longer be able to hold properly.

Lathe Attachments

Many attachments have been developed to increase the usefulness and the accuracy of the lathe. Here's a look at some of the most common. See Figure 5-3. Each will be discussed in more detail later. *Collets* are used to align and hold small workpieces. *Three-jaw* universal chucks have separate jaws for holding outside diameters, and for holding on inside diameters, are shown. The jaws of this chuck are actuated by the chuck key. They move in and out together and hold the round workpiece in a central location without the need for further adjustment. The jaws of a *four-jaw* independent chuck are adjusted independently of the other jaws by the chuck key. The jaws can be reversed in the chuck body in order to grip onto inside or outside surfaces. *Drill chucks* hold straightshank twist drills, reamers, and other cutting tools. It is usually held in the tailstock spindle when used in a lathe.

Figure 5-3 Here is a collection of drill chucks and three- and four-jaw chucks.
Courtesy of Sherline Products.

Cutting Tools

Single-point cutting tools are used on lathes, as well as shapers, and planers. Boring tools, which are also single-point cutting tools, are used on a large variety of power machine tools to locate and enlarge holes. The performance of these tools has a direct relation to the quality of the work done and the rate at which it is produced.

Cutting Tool Materials

Cutting tool materials must have the following properties at the high temperatures encountered in metal cutting: high hardness; good abrasion resistance; and resistance to chemical-metallurgical interaction with the work material. Cutting tools must also be able to withstand mechanical and thermal shock. The room temperature properties of cutting tool materials serve only as a rough guide to their ability to perform as a cutting tool; the only reliable guide is their actual performance when cutting.

High-Speed Steel Cutting Tools. High-speed steel is the name given to a group of very similar tool steels by virtue of their excellent red hardness. Red hardness is the ability of a steel to retain its hardness at a high temperature. High-speed steels will retain sufficient hardness at temperatures up to 1100°F to 1200°F which will enable them to cut other materials while at these temperatures. When cooled, the high-speed steels will return to their room temperature hardness. High-speed steels are very deep hardening; this enables them to be ground to a tool shape from solid stock and to be resharpened, when required, without a loss of hardness. See Figure 5-4. They can be softened by annealing and then machined into complex cutting tool shapes, such as twist drills, reamers, and milling cutters.

While the types of high-speed steels that are available are designed to meet some special requirement or to provide a special advantage, as a group they are almost a universal cutting tool material inasmuch as they are not too sensitive to the characteristic of the material being cut. Most work materials can be cut successfully with any type of high-speed steel.

Figure 5-4 Cutting tools made of high-speed steel can be sharpened repeatedly. This is a parting, or cut-off tool.

Photo by Tom Lipton

107

It is not necessary to be overly concerned about the type of high-speed steel used. Exceptions are the high-temperature alloys and other exotic materials. The limitations of high-speed steel relative to other cutting tool materials are its lower hardness and the slower cutting speeds that must be used.

Carbides. Cemented carbides, also called sintered carbides or just carbides, are much harder and more wear resistant than high-speed steel. They can retain their hardness at a higher temperature (1400°F and higher) than high-speed steel which enables them to cut at much faster cutting speeds. Like most very hard materials, carbides are somewhat brittle. Carbides are expensive; for this reason carbide cutting tools are made in the form of small tips brazed onto a steel shank, or small inserts that are mechanically clamped in place. See Figure 5-5.

Many different grades of cemented carbides are available and it is important to select the right grade. The cutting speeds for carbides given in the tables of this book are based on the assumption that the correct grade of carbide is being used. To select a carbide grade, it is best to seek the recommendation of a carbide producer. In general, select the carbide grade having the highest hardness with sufficient strength to prevent the cutting edge from chipping.

Parts of a Single-Point Cutting Tool

The nomenclature of a single-point cutting tool is shown in Figure 5-6. Most cutting tools have two cutting edges; a side-cutting edge and endcutting edge (see View A, Figure 5-6). When doing ordinary turning, the cutting is done with the side-cutting edge. The nose is that portion of the cutting edge which serves to connect the two cutting edges. It is a very critical part of the cutting edge because it produces the finished surface in turning. The flank of the tool is the surface on the side of the tool below the cutting edges. It must be relieved in order to permit the cutting edges to penetrate into the metal being cut. If the flanks below the side-cutting edge, nose, and

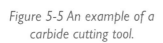

Figure 5-5 An example of a carbide cutting tool.

end-cutting edge were not relieved, the tool would rub and could not penetrate into the workpiece. The metal remaining in back of the cutting edge is the shank. It supports the cutting edge and provides a surface upon which the cutting tool can rest and be held.

The cutting tool in Figure 5-6 is a right-hand tool. A right-hand cutting tool can be identified by having the side-cutting edge on the left when viewed from the shank. A left-hand cutting tool has the side-cutting edge on the right side when viewed from the shank. The tool shown is also classified as a side-cutting-edge tool because it is provided with a side cutting edge and because it is intended that most of the cutting will be done with the side-cutting edge. End-cutting-edge tools (see H, J, and K, Figure 5-7) are intended to cut exclusively with their end-cutting edges and are made without a side-cutting edge.

Relief Angles. The side relief angle provides the relief for the side-cutting edge in order to allow it to penetrate into the workpiece so that it can form the chip. The end relief angle allows the end-cutting edge to penetrate into the workpiece. The relief angle below the nose is a blend of both the end relief and the side relief angles. Usually, both relief angles are equal although this is not always the case. The size of the relief angles has a very pronounced effect on the performance of the cutting tool; if they are too small they will cause a decided decrease in the tool life; if too large, the support below the cutting edge will be weakened to the extent that the cutting edge may break or chip. For average work, a relief angle of 10 degrees is recommended. For high-speed steel cutting tools the relief angle should be in the range of 8 to 16 degrees. Harder materials may require a smaller relief angle while softer materials are more successfully cut when a larger relief angle is used. The relief angle for carbide cutting tools should be in the range of 5 to 12 degrees. Since carbides are harder and more brittle than high-speed steel, a better support under the cutting edge is required.

Specifying a Single-Point Cutting Tool

Specify the width and height of the shank. In addition, it is necessary to specify the radius of the nose and the six principal tool angles which are:

1. End relief angle

2. Side relief angle

3. Back rake angle

4. Side rake angle

5. End-cutting-edge angle

6. Side-cutting-edge angle

These angles are all shown in View B, Figure 5-6 on page 92.

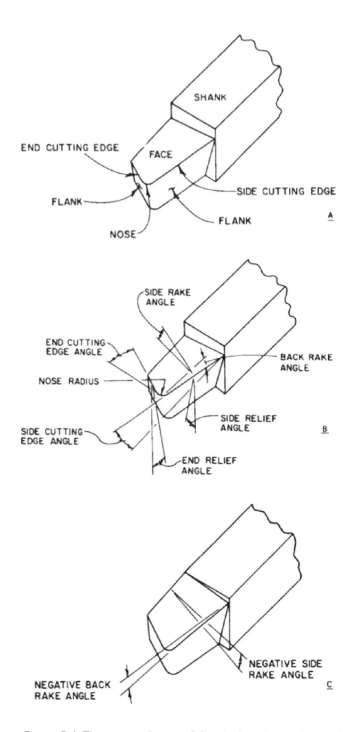

Figure 5-6 The nomenclature of the single-point cutting tool.

Rake Angles. The slope of the face of the cutting tool is determined by the back-rake and side-rake angles. See Tale 5-1. The side-rake angle is measured perpendicular to the side-cutting edge and the back-rake angle is measured parallel to the side-cutting edge. The rake angles may

Table 5-1 Recommended Rake Angles

Material	Hardness, HB	High-Speed Steel		Carbide	
		Back Rake Angle, deg.	Side Rake Angle, deg.	Back Rake Angle, deg.	Side Rake Angle, deg
Plain Carbon Steel	100 to 200	5 to 10	10 to 20	0 to 5	7 to 15
	200 to 300	5 to 7	8 to 12	0 to 5	5 to 8
	300 to 400	0 to 5	5 to JO	–5 to 0	3 to 5
	400 to 500	–5 to 0	–5 to 0	–8 to 0	–6 to 0
Alloy Steel	100 to 200	5 to 10	10 to 16	0 to 5	5 to 15
	200 to 300	5 to 7	8 to 12	0 to 5	5 to 8
	300 to 400	0 to 5	5 to 10	–5 to 0	3 to 5
	400 to 500	–5 to 0	–5 to 0	–6 to 0	–6 to 0
Aluminum					
Non Heat-Treated	...	10 to 20	30 to 35	0 to 15	15 to 30
Heat Treated	...	5 to 12	15 to 20	0 to 5	8 to 15
Magnesium	...	5 to 10	10 to 20	0 to 5	10 to 20
Stainless Steel					
Ferritie	130 to 190	5 to 7	8 to 10	0 to 5	5 to 7
Austenitic	130 to 190	5 to 7	8 to 12	0 to 5	5 to 7
Martensitic	130 to 220	0 to 5	5 to 8	–5 to 5	6 to 15
Gray Cast-Iron	100 to 200	5 to 10	10 to 15	0 to 5	6 to 15
	200 to 300	5 to 7	5 to 10	0 to 5	5 to 7
	300 to 400	–5 to 0	–5 to 0	–6 to 0	–6 to 0
Malleable Iron Ferritic Pearlitic	110 to 160	5 to 15	12 to 20	0 to 10	7 to 15
	160 to 200	5 to 8	10 to 12	0 to 5	5 to 8
	200 to 280	0 to 5	–5 to 8	–5 to 5	–5 to 8
Brass-Free Cutting	...	–5 to 5	0 to 10	...	...
Brass					
Red, Yellow, Naval,					
Manganese Bronze,					
Nickel Silver	..	–5 to 5	–5 to 5	–5 to 5	–5 to 5
Hard Phosphor Bronze	...	–5 to 0	–6 to 3	–5 to 0	–5 to 5

be positive, as shown at B in Figure 5-6, or may be negative, as shown at C. Note that the positive rake angles cause the face of the cutting tool to slant downward when moving away from the cutting edges, and that negative rake angles cause this surface to slant upward when moving away from the cutting edges. Positive side-rake angles are generally preferred because less cutting force is required in order to take a given size cut as compared to a tool with a negative side-rake angle.

End-Cutting-Edge Angle. The end-cutting-edge angle is the amount that the end-cutting edge slopes away from the nose of the tool, so that it will clear the finished surface on the workpiece when cutting with the side-cutting edge. The size of this angle is very important, particularly when cutting materials that tend to form a large crater on the face of the tool. This crater will then tend to enlarge toward the end-cutting edge where it will eventually break through and cause the tool to fail. When severe cratering occurs, the size of the end-cutting edge should be limited to 8 to 15 degrees; otherwise, it can be made as large as 45 degrees without an adverse effect on the performance of the tool.

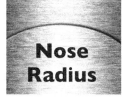

Nose Radius

The nose is an extension of the cutting edges. Its point must be formed into a radius in order to prevent the formation of a sharp threadlike groove on the surface of the work. If the nose radius is large, there will be a large area where a very thin chip is formed. When this occurs the cutting edge may fail to penetrate the work in this region, thus causing it to rub, especially if the cutting edge is dull. Should the cutting edge fail to penetrate the work, chatter may occur. Reducing the size of the nose radius will reduce the tendency to chatter. When finish turning, the nose radius must be made large enough to prevent any formation of threadlike grooves that have the form of the nose. If the size of the nose radius cannot be increased, the feed must be reduced to obtain a good surface finish. The size of the nose radius is then dependent upon the feed rate, the surface finish required on the work, and the requirement to prevent the occurrence of chatter—where either the tool or the workpiece vibrates leaving tool marks on the finished workpiece. It should be pointed out that the life of the cutting tool can be adversely affected by a nose radius that is either too large or too small. For average conditions when turning, a nose radius of 1/32 to 1/16 inch is recommended, although larger nose radii can sometimes be used successfully.

Tool Holders

In most cases the cutting tool is clamped in a tool holder; however, single-point cutting tools are sometimes clamped directly on the machine. The tool holder is made from a less expensive steel than the cutting tool. It is heat treated to increase its strength and to resist the penetration of the screws which clamp it in place. Carbide tips are sometimes brazed on steel shanks.

Most carbide cutting tools have several cutting edges and are called indexable inserts. When a cutting edge dulls, the insert is indexed, or moved, to bring another edge into cutting position on the tool holder. When all the cutting edges are used up, the insert is replaced.

High-Speed Steel Cutting Tools

High-speed steel single point cutting tools are usually ground to shape from solid blanks called tool bits. The basic high-speed steel cutting tool shapes and how they are applied are shown in Figure 5-7 and are described below:

A. Turning Tool. This is a basic high-speed steel single point cutting tool shape; it is a right-hand turning tool used for both rough and finish turning operations. The side cutting edge angle is zero degrees providing a zero degree lead angle when the tool is positioned as shown. It is often used to turn a cylindrical surface and to form a square shoulder at the end of the cut.

B. The Lead Angle Turning Tool. The lead angle (side cutting edge angle) turning tool is one of the most frequently used single point cutting tools for roughing and for finishing operations. It may be positioned at an angle, as shown at C, to form a square shoulder at the end of a turning cut.

C. Turning Tool with Chip Control. This tool is used to turn ductile materials, which form continuous chips that sometimes have a tendency to snarl. The chip control groove will curl the chip and cause it to break. There are several variations of this tool, one of which does not have the small flat on the face of the cutting edge. However, the flat is frequently used as it strengthens the cutting edge and tends to increase the tool life. It should be approximately .016 to .047 inch (0.40 to 1.2 mm) wide. The width of the flat and size of the groove depend on the material being cut and the feed rate used.

D. Round Nose Turning Tool. The round nose turning tool is used extensively when turning large diameter workpieces ; however, it can be used equally effectively on smaller diameter workpieces. The curved side cutting edge is, in effect, a large lead angle allowing a heavy feed rate to be used, while the large nose radius produces a very good surface finish on the workpiece. It is used for both heavy rough turning cuts and for light finish turning cuts.

continued on next page

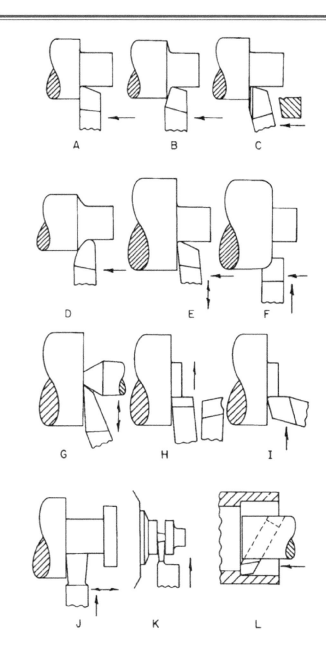

Figure 5-7. Typical tool shapes ground on high-speed tool bits and how they are used.

E. Turn and Facing Tool. This is merely a variation of the application of the turning tool at A, although for this purpose the tool is often provided with a slightly larger end cutting edge angle. Positioned as shown, it is used to turn a cylindrical surface up to a shoulder. When against the shoulder it is fed outward away from the workpiece center to take a facing cut—called back facing—in order to cut the shoulder perfectly square. If necessary, a facing cut can also be taken by feeding the tool inward toward the center of the workpiece. continued on next page

F. Form Tool. Form tools are used to turn a contoured shape on the workpiece, such as the radius in the illustration.

G. End Facing Tool. This tool is used to face the end of the workpiece while it is mounted on a tailstock center.

H. Facing Tool. The facing tool shown is used to take heavy facing cuts requiring a large amount of stock removal. It has a "hook" ground on its face to provide a rake angle to the end cutting edge. Sometimes it is provided with a narrow flat on the cutting edge, similar to the flat on the tool shown at C. The side flank adjacent to the shoulder has a 5- to 8-degree relief angle.

I. Facing Tool. Actually this is a left-hand turning tool which when used as shown is sometimes called a facing tool. It is used to take heavy or light facing cuts. Having a nose radius it will usually produce a better surface finish than the tool at H.

J. Grooving or Necking Tool. The end cutting edge is the primary cutting edge of this tool; the sides have a slight back clearance angle to prevent their rubbing against the sides of the groove. Wide grooves are cut by taking several plunge cuts into the workpiece. The groove is then finished by taking light finishing cuts on the cylindrical surface using the carriage feed and on the sides using the cross-slide feed. Narrow grooves are cut by a single plunge into the workpiece, with the width of the tool being made equal to the width of the groove in the workpiece.

K. Cut-Off Tool. Cut-off tools, also called parting tools, are used to cut off the ends of stock in a lathe. Usually the portion that is cut off is a partially or completely finished workpiece.

L. Boring Tool. Boring tools are used to enlarge holes. There are many different styles of boring tools of which the tool illustrated is one example. Three features of these tools must be carefully controlled: 1. the end relief angle must be large enough to provide an adequate clearance with respect to the wall of the hole; 2. the nose radius should be as small as possible; and, 3. the lead angle should he small, preferably zero degree for small diameter boring bars. The nose radius and the lead angle tend to deflect the boring bar away from the wall of the hole. They should, therefore, be kept as small as possible.

Carbide Cutting Tools

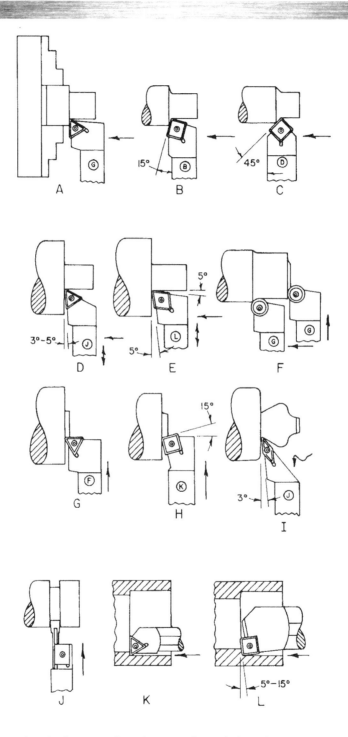

Figure 5-8 Typical tools shapes and applications for carbide, indexable-insert cutting tools. Circled letters are the standard tool holder designations.

Figure 5-8 on the previous page illustrates a group of typical carbide indexable insert cutting tools and how they are applied. When selecting negative rake indexable insert cutting tools, it is necessary to know where the relief angles are located since the tool can only cut with an edge that has a relief angle. The relief angle on these inserts is provided by holding them at a compound angle on the tool holder. The arrows on the illustrations show the direction in which negative rake tools will cut. Since the relief angles are ground or formed on positive rake inserts, they are not a cause of concern if the tool holder is correctly made.

A. Turning. The G-style tool holder illustrated has an offset to allow the triangular insert to cut close to chuck jaws or to a large shoulder. An A style tool holder is similar but does not have the offset. These tool holders have a zero degree lead angle and can be used to form a square shoulder following a turning operation.

B. Turning. The B-style tool holder provides a 15-degree lead angle and allows a square insert to be used, which has more cutting edges available than the triangular insert.

C. Turning. The D-style tool holder provides a 45-degree lead which allows a heavier feed rate to be used.

D. Turn and Backface. This operation is used to produce shoulders that are precise and square. The J-style tool holder positions the insert to have a negative lead angle; i.e., it will cut with the nose leading. The procedure used is to turn the cylindrical surface to the shoulder, and then to back-face the shoulder. If necessary, this tool can also be used to take a light facing cut by feeding inward toward the center of the workpiece.

E. Turn and Backface. In all respects the operation of this tool is identical to the tool shown at D. The insert is diamond shaped having an acute angle of 80 degree. When held on the L-style tool holder it has only two cutting edges per face available. However, this holder holds the insert very firmly and allows a heavier turning cut to be taken than is usually possible with the tool at D. The size of the back-facing cut is dependent primarily on the negative lead angle, although on tough materials the L-style holder will hold the insert more firmly when backfacing.

F. Turn and Facing. Round inserts provide more usable cutting edge, or indexes, per face before they are used up than the other insert shapes. Furthermore, the large radius produces a good surface finish. When used on the G-style tool holder, the round inserts can be used to take medium and light turning and facing cuts.

G. Facing. The F-style tool holder and the triangular insert are used to take light and heavy facing cuts.

H. Facing. The K-style tool holder utilizes a square insert to take light and heavy facing cuts.

continued on next page

I. Contour Turning. The diamond insert shown is specifically designed for contour turning, although other inserts are also used for this purpose depending on the shape of the contour. There are two types of diamond inserts designed for this purpose: one has a 35-degree acute angle while the other has a 55-degree acute angle. A J-style tool holder is shown; other style tool holders are available which hold the insert in a different position.

J. Cut-Off and Grooving. Cut-off and grooving tool holders do not have an ANSI designation. Many different designs of carbide cut-off and grooving tools are available.

K. Boring. A large selection of indexable insert boring bars are available. When the lead angle is zero degrees a triangular insert is used, as shown.

L. Boring. Boring bars having a lead angle can utilize square inserts. The lead angle should be kept small on smaller diameter bars.

Grinding Single-Point Cutting Tools

Single-point cutting tools are generally ground by hand on pedestal or bench-type grinders. Some of these are specifically designed to grind single-point cutting tools and have a table on which the tools can be placed while they are held against the grinding wheel by hand. The grinding wheel used on these machines is designed for grinding on the side of the wheel in order to produce a true plane surface on the workpiece. The table can be set at an angle, with respect to the side of the wheel, thereby enabling the desired angle to be ground on the tool. Provisions are usually made to have a flow of coolant available in order to prevent the tool from overheating while grinding. Most single-point, carbide cutting tools are ground on this type of grinder using a diamond-impregnated grinding wheel. High-speed steel cutting tools can also be ground on these machines using an aluminum oxide grinding wheel.

Figure 5-9 General-purpose, high-speed steel turning tool.

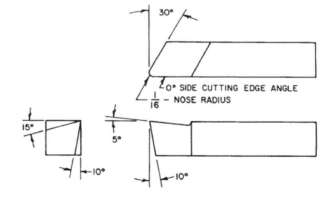

Most high-speed steel cutting tools, however, are ground on a simple pedestal or bench grinder. The tool bit is held by hand against the peripheral face of the wheel, at the angle required. This method is called offhand grinding, and it requires a considerable amount of skill. This skill can be acquired by analyzing what is to be accomplished and by practice.

A general-purpose, high-speed steel turning tool is shown in Figure 5-9. When grinding, the tool bit should be held with the cutting edge facing upward so that it can be seen. Held in this manner, the grinding wheel will "run on" to the cutting edge instead of running off from it. This prevents the possibility of chipping and reduces the tendency to form a fine almost microscopic feather on the cutting edge. All of the angles on the tool should first be rough ground almost up to the cutting edges; then the angles are finish ground and the cutting edges sharpened. The nose radius is ground last. It is ground by holding the tool at the correct angle with a light pressure and then swinging the tool bit slowly back and forth between the side and end flanks. See Figure 5-10.

What do Metal Cutting Fluids Do?

The functions of metal cutting fluids are:

1. Cool the cutting tool as a primary function, and also to cool the workpiece
2. Reduce the heat generated by friction through lubrication
3. Provide anti-weld properties in order to prevent welding of the chip to the tool
4. Wash away the chips

Figure 5-10 Here is a collection of special tool bit profiles.
Photo Tom Lipton

Metal Cutting Fluids

There are three basic types of cutting fluids: water soluble oils; straight cutting oils; and chemical fluids.

Water Soluble Oil. Also called emulsified oils, water soluble oils are mixed with water in a water-to-oil ratio ranging from 5 to 1 to as much as 50 to l. The mixture has an appearance ranging from milky to cloudy, or almost clear. Water is the best coolant, but has many disadvantages as a cutting fluid, the primary one being that it causes rust. Soluble oils prevent rust, are excellent coolants, and have a measure of lubricity. Additives enhance their performance and prevent bacterial growth. Soluble oils can be used for almost all light, medium, and heavy-duty machining operations. They can be used over the entire range of cutting speeds, including the higher speeds where carbide tools are used. They are not recommended for certain severe cutting operations requiring greater lubricity and anti-weld properties. Soluble oils are also recommended for almost all precision grinding operations.

Straight Cutting Oils. These are made from mineral oil that is usually compounded with additives to increase their effectiveness. Uncompounded mineral oils are sometimes used on very light duty machining operations on aluminum, magnesium, brass, and sulfurized free cutting steels. Straight oils should he used when the cutting speed is moderate or low and on work where soluble oils do not give a satisfactory performance. They are particularly recommended on jobs where the chip is crowded, such as thread cutting. Straight oils are used on some precision grinding operations; however, provisions must he made to prevent the vapors from exploding.

Chemical Fluids. Also called synthetic cutting fluids, chemical fluids are proprietary chemical compounds to which water is added. There are two kinds: those having a wetting agent and those that do not. Those that do not have a wetting agent are restricted to rough grinding operations, where they tend to keep the grinding wheel open and free cutting. Those having a wetting agent have some lubricity to allow the machine tool slides and other machine parts to function smoothly. Chemical fluids are excellent coolants and are recommended for machining operations that are in the moderate to higher cutting speed range.

Applying Cutting Fluids

In general, cutting fluids should be applied in a continuous stream that floods the cutting tool. An intermittent supply on a hot cutting tool will likely cause thermal cracks to form on the tool. It is not, however, poor practice to apply a cutting fluid manually with a brush or a squirt can whenever the cutting speed is moderate or low. There are many occasions in the shop where this practice is the most practical method of application. At these speeds the tool is not likely to be damaged and the application of the cutting fluid is beneficial.

The only real measure of a cutting fluid is how it performs on the job. When selecting a cutting fluid, the cutting speed should be considered. In general, when the cutting speed exceeds about 75 to 100 fpm (25 to 30 m/min) a water soluble oil or a chemical fluid should be used. Below this speed range, first consider water soluble oil; then if the job is too difficult for a water soluble oil, select a straight cutting oil. See Figure 5-11.

Figure 5-11 Cutting fluid applied to a workpiece.

Cutting Speeds

The cutting speed is given in terms of feet per minute (fpm). Sometimes this is called surface feet per minute (sfpm). In the metric system of units the cutting speed is specified in meters per minute (m/min). Tables 5-2 through 5-6 give the recommended cutting speeds for cutting many materials in terms of feet per minute. These cutting speeds may provide more information than the typical hobbyist may need. It is best to consider the tables a starting point, which may have to be modified by actual experience, but you can use the tables to help determine the correct rpm for the cutting speed chosen by plugging the information, along with the diameter of the workpiece, into the "Calculating Cutting Speed" equation found on page 129. Many factors affect the cutting speed which may combine to warrant an increase or sometimes a decrease in the cutting speed.

Tool Material. The type of material used to make the cutting tool influences the cutting speed. The tables list the cutting speeds for use with high-speed steel. Although there are slight differences in the cutting speed that can be used with tools made of different types of high-speed steel, these differences are generally small. A greater difference exists between different grades of carbides, and it is essential that the correct grade be used. Coated carbides cannot be used to cut all materials; however, where they can be used, they can usually be operated at a 20 to 40 percent faster than the speeds recommended by some industrial sources and sometimes by as much as 50 percent. Since there are so many grades of carbides, the selection of a specific grade should be made by obtaining the recommendations of a carbide producer.

Table 5-2. Recommended Cutting Speeds in Feet per Minute for Turning Plain Carbon and Alloy Steels. See the Turning Column.

Material AISI and SAE Steels	Hardness HB[a]	Material Condition	Cutting Speed, fpm HSS			
			Turning	Milling	Drilling	Reaming
Free Machining Plain Carbon Steels (Resulfurized)						
1212, 1213, 1215	100-150	HR, A	150	140	120	80
	150-200	CD	160	130	125	80
1108, 1109, 1115, 1117, 1118, 1120, 1126, 1211	100-150	HR, A	130	130	110	75
	150-200	CD	120	115	120	80
1132, 1137, 1139, 1140, 1144, 1146, 1151	175-225	HR, A, N, CD	120	115	100	65
	275-325	Q and T	75	70	70	45
	325-375	Q and T	50	45	45	30
	375-425	Q and T	40	35	35	20
Free Machining Plain Carbon Steels (Leaded)						
11L17, 11L18, 12L13, 12L14	100-150	HR, A, N, CD	140	140	130	85
	150-200	HR, A, N, CD	145	130	120	80
	200-250	N, CD	110	110	90	60
Plain Carbon Steels						
1006, 1008, 1009, 1010, 1012, 1015, 1016, 1017, 1018, 1019, 1020, 1021, 1022, 1023, 1024, 1025, 1026, 1513, 1514	100-125	HR, A, N, CD	120	110	100	65
	125-175	HR, A, N, CD	110	110	90	60
	175-225	HR, N, CD	90	90	70	45
	225-275	CD	70	65	60	40
1027, 1030, 1033, 1035, 1036, 1037, 1038, 1039, 1040, 1041, 1042, 1043, 1045, 1046, 1048, 1049, 1050, 1052, 1152, 1524, 1526, 1527, 1541	125-175	HR, A, N, CD	100	100	90	60
	175-225	HR, A, N, CD	85	85	75	50
	225-275	N, CD, Q and T	70	70	60	40
	275-325	Q and T	60	55	50	30
	325-375	Q and T	40	35	35	20
	375-425	Q and T	30	25	25	15
1055, 1060, 1064, 1065, 1070, 1074, 1078, 1080, 1084, 1086, 1090, 1095, 1548, 1551, 1552, 1561, 1566	125-175	HR, A, N, CD	100	90	85	55
	175-225	HR, A, N, CD	80	75	70	45
	225-275	N, CD, Q and T	65	60	50	30
	275-325	Q and T	50	45	40	25
	325-375	Q and T	35	30	30	20
	375-425	Q and T	30	15	15	10
Free Machining Alloy Steels (Resulfurized)						
4140, 4150	175-200	HR, A, N, CD	110	100	90	60
	200-250	HR, N, CD	90	90	80	50
	250-300	Q and T	65	60	55	30
	300-375	Q and T	50	45	40	25
	375-425	Q and T	40	35	30	15
Free Machining Alloy Steels (Leaded)						
41L30, 41L40, 41L47, 41L50, 43L47, 51L32, 52L100, 86L20, 86L40	150-200	HR, A, N, CD	120	115	100	65
	200-250	HR, N, CD	100	95	90	60
	250-300	Q and T	75	70	65	40
	300-375	Q and T	55	50	45	30
	375-425	Q and T	50	40	30	15

continued

Table 5-2. Recommended Cutting Speeds in Feet per Minute for Turning Plain Carbon and Alloy Steels. See the Turning Column. (*continued*)

Material AISI and SAE Steels	Hardness HB[a]	Material Condition	Cutting Speed, fpm HSS			
			Turning	Milling	Drilling	Reaming
Alloy Steels						
4012, 4023, 4024, 4028, 4118, 4320, 4419, 4422, 4427, 4615, 4620, 4621, 4626, 4718, 4720, 4815, 4817, 4820, 5015, 5117, 5120, 6118, 8115, 8615, 8617, 8620, 8622, 8625, 8627, 8720, 8822, 94B17	125-175	HR, A, N, CD	100	100	85	55
	175-225	HR, A, N, CD	90	90	70	45
	225-275	CD, N, Q and T	70	60	55	35
	275-325	Q and T	60	50	50	30
	325-375	Q and T	50	40	35	25
	375-425	Q and T	35	25	25	15
1330, 1335, 1340, 1345, 4032, 4037, 4042, 4047, 4130, 4135, 4137, 4140, 4142, 4145, 4147, 4150, 4161, 4337, 4340, 50B44, 50B46, 50B50, 50B60, 5130, 5132, 5140, 5145, 5147, 5150, 5160, 51B60, 6150, 81B45, 8630, 8635, 8637, 8640, 8642, 8645, 8650, 8655, 8660, 8740, 9254, 9255, 9260, 9262, 94B30	175-225	HR, A, N, CD	85	75	75	50
	225-275	N, CD, Q and T	70	60	60	40
	275-325	N, Q and T	60	50	45	30
	325-375	N, Q and T	40	35	30	15
	375-425	Q and T	30	20	20	15
E51100, E52100	175-225	HR, A, CD	70	65	60	40
	225-275	N, CD, Q and T	65	60	50	30
	275-325	N, Q and T	50	40	35	25
	325-375	N, Q and T	30	30	30	20
	375-425	Q and T	20	20	20	10
Ultra High Strength Steels (Not AISI)						
AMS 6421 (98B37 Mod.), AMS 6422 (98BV40), AMS 6424, AMS 6427, AMS 6428, AMS 6430, AMS 6432, AMS 6433, AMS 6434, AMS 6436, AMS 6442, 300M, D6ac	220-300	A	65	60	50	30
	300-350	N	50	45	35	20
	350-400	N	35	20	20	10
	43-48 HRC	Q and T	25	…	…	…
	48-52 HRC	Q and T	10	…	…	…
Maraging Steels (Not AISI)						
18% Ni Grade 200, 18% Ni Grade 250, 18% Ni Grade 300, 18% Ni Grade 350	250-325	A	60	50	50	30
	50-52 HRC	Maraged	10	…	…	…
Nitriding Steels (Not AISI)						
Nitralloy 125, Nitralloy 135, Nitralloy 135 Mod., Nitralloy 225, Nitralloy 230, Nitralloy N, Nitralloy EZ, Nitrex I	200-250	A	70	60	60	40
	300-350	N, Q and T	30	25	35	20

[a] Abbreviations designate: HR, hot rolled; CD, cold drawn; A, annealed; N, normalized; Q and T, quenched and tempered; and HB, Brinell hardness number.

Speeds for turning based on a feed rate of 0.012 inch per revolution and a depth of cut of 0.125 inch.

Table 5-3. Recommended Cutting Speeds in Feet per Minute for Turning Tool Steels. See the Turning Column.

Material Tool Steels (AISI Types)	Hardness HB[a]	Material Condition	Cutting Speed, fpm HSS			
			Turning	Milling	Drilling	Reaming
Water Hardening W1, W2, W5	150-200	A	100	85	85	55
Shock Resisting S1, S2, S5, S6, S7	175-225	A	70	55	50	35
Cold Work, Oil Hardening O1, O2, O6, O7	175-225	A	70	50	45	30
Cold Work, High Carbon High Chromium D2, D3, D4, D5, D7	200-250	A	45	40	30	20
Cold Work, Air Hardening A2, A3, A8, A9, A10	200-250	A	70	50	50	35
A4, A6	200-250	A	55	45	45	30
A7	225-275	A	45	40	30	20
Hot Work, Chromium Type H10, H11, H12, H13, H14, H19	150-200	A	80	60	60	40
	200-250	A	65	50	50	30
	325-375	Q and T	50	30	30	20
	48-50 HRC	Q and T	20	…	…	…
	50-52 HRC	Q and T	10	…	…	…
	52-54 HRC	Q and T	…	…	…	…
	54-56 HRC	Q and T	…	…	…	…
Hot Work, Tungsten Type H21, H22, H23, H24, H25, H26	150-200	A	60	55	55	35
	200-250	A	50	45	40	25
Hot Work, Molybdenum Type H41, H42, H43	150-200	A	55	55	45	30
	200-250	A	45	45	35	20
Special Purpose, Low Alloy L2, L3, L6	150-200	A	75	65	60	40
Mold P2, P3, P4, P5, P6	100-150	A	90	75	75	50
P20, P21	150-200	A	80	60	60	40
High Speed Steel M1, M2, M6, M10, T1, T2, T6	200-250	A	65	50	45	30
M3-1, M4, M7, M30, M33, M34, M36, M41, M42, M43, M44, M46, M47, T5, T8	225-275	A	55	40	35	20
T15, M3-2	225-275	A	45	30	25	15

[a] Abbreviations designate: A, annealed; Q and T, quenched and tempered; HB, Brinell hardness number; and HRC, Rockwell C scale hardness number.

Speeds for turning based on a feed rate of 0.012 inch per revolution and a depth of cut of 0.125 inch.

124

Table 5-4. Recommended Cutting Speeds in Feet per Minute for Turning Stainless Steels. See the Turning Column.

Material	Hardness HB[a]	Material Condition	Cutting Speed, fpm HSS			
			Turning	Milling	Drilling	Reaming
Free Machining Stainless Steels (Ferritic)						
430F, 430F Se	135-185	A	110	95	90	60
(Austenitic), 203EZ, 303, 303Se, 303MA, 303Pb, 303Cu, 303 Plus X	135-185	A	100	90	85	55
	225-275	CD	80	75	70	45
(Martensitic), 416, 416Se, 416Plus X, 420F, 420FSe, 440F, 440FSe	135-185	A	110	95	90	60
	185-240	A,CD	100	80	70	45
	275-325	Q and T	60	50	40	25
	375-425	Q and T	30	20	20	10
Stainless Steels						
(Ferritic), 405, 409, 429, 430, 434, 436, 442, 446, 502	135-185	A	90	75	65	45
(Austenitic), 201, 202, 301, 302, 304, 304L, 305, 308, 321, 347, 348	135-185	A	75	60	55	35
	225-275	CD	65	50	50	30
(Austenitic), 302B, 309, 309S, 310, 310S, 314, 316, 316L, 317, 330	135-185	A	70	50	50	30
(Martensitic), 403, 410, 420, 501	135-175	A	95	75	75	50
	175-225	A	85	65	65	45
	275-325	Q and T	55	40	40	25
	375-425	Q and T	35	25	25	15
(Martensitic), 414, 431, Greek Ascoloy	225-275	A	60	55	50	30
	275-325	Q and T	50	45	40	25
	375-425	Q and T	30	25	25	15
(Martensitic), 440A, 440B, 440C	225-275	A	55	50	45	30
	275-325	Q and T	45	40	40	25
	375-425	Q and T	30	20	20	10
(Precipitation Hardening), 15-5PH, 17-4PH, 17-7PH, AF-71, 17-14CuMo, AFC-77, AM-350, AM-355, AM-362, Custom 455, HNM, PH13-8, PH14-8Mo, PH15-7Mo, Stainless W	150-200	A	60	60	50	30
	275-325	H	50	50	45	25
	325-375	H	40	40	35	20
	375-450	H	25	25	20	10

[a] Abbreviations designate: A, annealed; CD, cold drawn: N, normalized; H, precipitation hardened; Q and T, quenched and tempered; and HB, Brinell hardness number.

Speeds for turning based on a feed rate of 0.012 inch per revolution and a depth of cut of 0.125 inch.

Table 5-5. Recommended Cutting Speeds in Feet per Minute for Turning Ferrous Cast Metals. See the Turning Column.

Material	Hard-ness HB[a]	Material Condition	Cutting Speed, fpm HSS			
			Turning	Milling	Drilling	Reaming
Gray Cast Iron						
ASTM Class 20	120-150	A	120	100	100	65
ASTM Class 25	160-200	AC	90	80	90	60
ASTM Class 30, 35, and 40	190-220	AC	80	70	80	55
ASTM Class 45 and 50	220-260	AC	60	50	60	40
ASTM Class 55 and 60	250-320	AC, HT	35	30	30	20
ASTM Type 1, 1b, 5 (Ni Resist)	100-215	AC	70	50	50	30
ASTM Type 2, 3, 6 (Ni Resist)	120-175	AC	65	40	40	25
ASTM Type 2b, 4 (Ni Resist)	150-250	AC	50	30	30	20
Malleable Iron						
(Ferritic), 32510, 35018	110-160	MHT	130	110	110	75
(Pearlitic), 40010, 43010, 45006, 45008, 48005, 50005	160-200	MHT	95	80	80	55
	200-240	MHT	75	65	70	45
(Martensitic), 53004, 60003, 60004	200-255	MHT	70	55	55	35
(Martensitic), 70002, 70003	220-260	MHT	60	50	50	30
(Martensitic), 80002	240-280	MHT	50	45	45	30
(Martensitic), 90001	250-320	MHT	30	25	25	15
Nodular (Ductile) Iron						
(Ferritic), 60-40-18, 65-45-12	140-190	A	100	75	100	65
(Ferritic-Pearlitic), 80-55-06	190-225	AC	80	60	70	45
	225-260	AC	65	50	50	30
(Pearlitic-Martensitic), 100-70-03	240-300	HT	45	40	40	25
(Martensitic), 120-90-02	270-330	HT	30	25	25	15
	330-400	HT	15	–	10	5
Cast Steels						
(Low Carbon), 1010, 1020	100-150	AC, A, N	110	100	100	65
(Medium Carbon), 1030, 1040, 1050	125-175	AC, A, N	100	95	90	60
	175-225	AC, A, N	90	80	70	45
	225-300	AC, HT	70	60	55	35
(Low Carbon Alloy), 1320, 2315, 2320, 4110, 4120, 4320, 8020, 8620	150-200	AC, A, N	90	85	75	50
	200-250	AC, A, N	80	75	65	40
	250-300	AC, HT	60	50	50	30
(Medium Carbon Alloy), 1330, 1340, 2325, 2330, 4125, 4130, 4140, 4330, 4340, 8030, 80B30, 8040, 8430, 8440, 8630, 8640, 9525, 9530, 9535	175-225	AC, A, N	80	70	70	45
	225-250	AC, A, N	70	65	60	35
	250-300	AC, HT	55	50	45	30
	300-350	AC, HT	45	30	30	20
	350-400	HT	30	…	20	10

[a] Abbreviations designate: A, annealed; AC, as cast; N, normalized; HT, heat treated; MHT, malleabilizing heat treatment; and HB, Brinell hardness number.

Speeds for turning based on a feed rate of 0.012 inch per revolution and a depth of cut of 0.125 inch.

Table 5-6 Recommended Cutting Speeds in Feet per Minute for Turning, Milling, Drilling, and Reaming Light Metals. See the Turning Column.

Material Light Metals	Material Condition[a]	Cutting Speed, fpm HSS			
		Turning	Milling	Drilling	Reaming
All Wrought Aluminum Alloys	CD	600	600	400	400
	ST and A	500	500	350	350
All Aluminum Sand and Perma-nent Mold Casting Alloys	AC	750	750	500	500
	ST and A	600	600	350	350
All Aluminum Die Casting Alloys	AC	125	125	300	300
	ST and A	100	100	70	70
except Alloys 390.0 and 392.0	AC	80	80	125	100
	ST nd A a	60	60	45	40
All Wrought Magnesium Alloys	A, CD, ST, and A	800	800	500	500
All Cast Magnesium Alloys	A, AC, ST, and A	800	800	450	450

[a] Abbreviations designate: A, annealed; AC, as cast; CD, cold drawn; ST and A, solution treated as aged.

Work Material. Of equal importance in selecting the cutting speed is the work material. Each work material will have a range of cutting speeds at which it can be cut. The cutting speeds recommended in this chapter are based on the expectancy that the material can be cut at the given speed with a reasonable tool life. For most materials, the cutting speeds in the tables are given for several different hardness levels. Since the hardness is not always known, the material condition associated with the hardness is provided in a separate column. When in doubt, start at the lower cutting speed and then see if the workpiece can be cut faster.

Tool Life. Tool life is the length of time that a cutting tool will cut before it becomes dull or requires replacement. When cutting most metals, cutting tools will operate successfully over a rather wide range of cutting speeds, up to a cutting speed where the tool life is too short to be acceptable. Below this speed the tools will behave in the following manner when normal flank wear occurs and an abnormal type of tool failure does not occur.

A reduction in the cutting speed will cause a much larger increase in the tool life. Conversely, increasing the cutting speed will cause a much larger· reduction in the tool life.

Cutting tools can perform successfully over a rather wide range of cutting speeds except when cutting very hard materials and certain other difficult-to-machine materials. The cutting speed selected should be based on a desired tool life. If the cutting speed used is found to result in a tool life that is too short, the first remedial step should be to reduce the cutting speed. While a slow cutting speed will result in a very long tool life, this is always at the expense of an unnecessarily long cutting time and an increased part cost.

Feed and Depth of Cut. The feed rate and the depth of cut also have an effect on the tool life and, therefore, on the cutting speed. The cutting speeds in Tables 5-2 through 5-6 are based on using a feed rate of .012 in./rev (0.30 mm/rev) and a depth of cut of .125 in. (3.18 mm). This is an aggressive cut, and probably one that you will not make in your home shop. You can make the adjustment by using the factors for feed and depth of cut found in Table 5-7. Note: don't use the the factors in 5-7 with Table 5-6 because their validity for many nonferrous metals is uncertain. Applying the feed and depth of cut factors will show that when light cuts are taken the cutting speed can be increased and when heavy roughing cuts are taken it should be decreased.

Cutting Tool Geometry. This topic is treated at length elsewhere in this chapter. It is, therefore sufficient to say here that the shape of the cutting tool has some effect on the cutting speed.

Cutting Fluids. Coolant type cutting fluids, when correctly applied, will lower the temperature of the cutting tool and tend to increase the tool life. As an alternative, the coolant will cool the tool enough so that a higher cutting speed can be used without overheating. The degree of improvement that can be obtained depends on the type of coolant used.

Selecting the Cutting Conditions

The tool life of a cutting tool is most affected by the cutting speed, secondly by the feed rate, and least by the depth of cut. This important principle of metal cutting should be memorized. Stated in another way, reducing the cutting speed will increase the tool life more than a reduction in the feed rate; and a reduction in the feed rate will increase the tool life more than a reduction in the depth of cut. Conversely, increasing the depth of cut will reduce the tool life less than an increase in the feed rate; and, increasing the feed rate will reduce the tool life less than an increase in the cutting speed. The latter principle leads to the logical sequence of selecting the cutting conditions, which are given below:

1. Select the depth of cut. Use the largest depth of cut possible, as determined by the amount of metal to be removed from the part and by the available power on the machine.

2. Select the feed rate. Use the heaviest, feed rate possible consistent with the surface finish requirement on the workpiece, the rigidity of the machine and workpiece, and the available power on the machine.

3. Select the cutting speed. Use an appropriate table of cutting speeds as a guide, such as provided in this book.

4. Find the appropriate feed and depth of cut factors, when available. Use Table 5-7 on the following page, but only ·with the cutting speed tables provided in this book and in Machinery's Handbook. Do not use this table for nonferrous metals and alloys.

5. Calculate the cutting speed and machine spindle speed. See below

Calculating the Cutting Speed

The formulas used to calculate the cutting speed and the spindle speed·of the lathe are given below for both customary inch and metric units. Either inch or metric units can be used in Equation 5-1. The answer will be in terms of the units used for the cutting speed selected from the table.

$$V = V_0 F_f F_d \tag{5-1}$$

$$N = \frac{12 V}{\pi D} \tag{Inch units only) (5-2}$$

$$N = \frac{1000 V}{\pi D} \tag{Metric units only) (5-3}$$

Where:

V = Cutting speed as modified for feed and depth of cut

V_0 = Cutting speed from Tables 5-2 through 5-5, fpm or m/min

F_f = Feed factor, from Table 5-7

F_d = Depth of cut factor, from Table 5-7

N = Spindle speed of lathe, rpm

D = Diameter of workpiece, inches or mm. (This is the diameter before turning, not the finish turned diameter.)

π = 3.14 (pi)

Example 5-1:

Calculate the cutting speed and the spindle speed for turning a 1.500-inch- diameter bar of AISI 4340 steel having a hardness of 220 HB using a high-speed steel cutting tool. The depth of cut selected is .200 in. and the feed selected is .020 in./rev.

V_0 = 70 fpm (from Table 5-2)

F_f = .80 (from Table 5-7)

$$V = V_0 F_f F_d = 70 \times 0.80 \times 0.93$$
$$= 52 \text{ fpm}$$

$$N = \frac{12V}{\pi D} = \frac{12 \times 52}{\pi \times 1.500}$$
$$= 125 \text{ rpm}$$

In calculations of this type the answers are always rounded-off (e.g., the calculated cutting speed is 52.08 fpm, which is rounded-off to 52 fpm). The actual spindle speed that would be used is the closet available on the lathe. Sometimes, but not always, when the closest available speed is much faster than the calculated spindle speed, the closest lower spindle speed is used.

Table 5-7. Cutting Speed Feed and Depth of Cut Factors for Turning*

Feed		Feed Factor	Depth of Cut		Depth-of-Cut Factor
in.	mm	F_f	in.	mm	F_d
0.002	0.05	1.50	0.005	0.13	1.50
0.003	0.08	1.50	0.010	0.25	1.42
0.004	0.10	1.50	0.016	0.41	1.33
0.005	0.13	1.44	0.031	0.79	1.21
0.006	0.15	1.34	0.047	1.19	1.15
0.007	0.18	1.25	0.062	1.57	1.10
0.008	0.20	1.18	0.078	1.98	1.07
0.009	0.23	1.12	0.094	2.39	1.04
0.010	0.25	1.08	0.100	2.54	1.03
0.011	0.28	1.04	0.125	3.18	1.00
0.012	0.30	1.00	0.150	3.81	0.97
0.013	0.33	0.97	0.188	4.78	0.94
0.014	0.36	0.94	0.200	5.08	0.93
0.015	0.38	0.91	0.250	6.35	0.91
0.016	0.41	0.88	0.312	7.92	0.88
0.018	0.46	0.84	0.375	9.53	0.86
0.020	0.51	0.80	0.438	11.13	0.84
0.022	0.56	0.77	0.500	12.70	0.82
0.025	0.64	0.73	0.625	15.88	0.80
0.028	0.71	0.70	0.688	17.48	0.78
0.030	0.76	0.68	0.750	19.05	0.77
0.032	0.81	0.66	0.812	20.62	0.76
0.035	0.89	0.64	0.938	23.83	0.75
0.040	1.02	0.60	1.000	25.40	0.74
0.045	1.14	0.57	1.250	31.75	0.73
0.050	1.27	0.55	1.250	31.75	0.72
0.060	1.52	0.50	1.375	34.93	0.71

* For use with HSS tool data only from Tables 1 through 8. Adjusted cutting speed $V = V_{HSS} \times F_f \times F_d$, where V_{HSS} is the tabular speed for turning with high-speed tools.

.

6 Working on a Lathe

The basic operations performed on an engine lathe are shown in Figure 6-1. Those operations performed on external surfaces with a single point cutting tool are called turning. In general, all machining operations, including turning and boring, can be classified as either roughing or finishing. The objective of a roughing operation is to remove the bulk of the material as rapidly and as efficiently as possible, while leaving a small amount of material on the workpiece for the finishing operation. Finishing operations are performed to obtain the final size, shape, and surface finish.

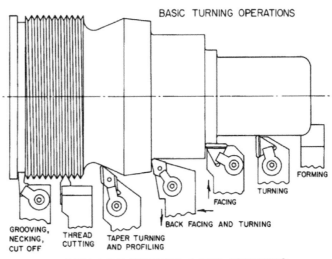

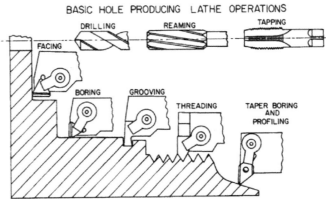

Figure 6-1 The basic turning operations performed on a lathe. (Top) On outside diameters. (Bottom) In producing holes.

In turning work, the lathe rotates the workpiece while a stationary cutting tools shapes the metal. For precision's sake, the workpiece must be firmly supported. There are a number of ways to support the work, including turning between centers and using chucks and collets.

Turning Between Lathe Centers

Generally, longer workpieces are turned while supported on one or two lathe centers—components attached to the headstock and tailstock that hold the work in place. Cone shaped holes, called center holes, which fit the lathe centers are drilled in the ends of the workpiece. The end of the workpiece adjacent to the tailstock is always supported by a tailstock center, while the end near the headstock may be supported by a headstock center.

Very precise results can be obtained by supporting the workpiece between two centers. One end of the workpiece is machined; then the workpiece can be turned around in the lathe to machine the other end. The center holes in the workpiece serve as precise locators as well as bearing surfaces to carry the weight of the workpiece and to resist the cutting forces. When the workpiece has been removed from the lathe, the center holes will accurately align the workpiece later, either in the original lathe or a different one.

Preparing the Workpiece

The first step is to drill holes in the ends of the workpiece. These holes are drilled with a combination drill and countersink which is commonly called a center drill (see Figure 6-2). The conical hole should be drilled deep enough so that a good bearing surface is provided for the lathe centers—the components that hold the work in the lathe. The holes must not, however, be drilled beyond the largest diameter of the center drill. The countersink has a 60-degree included angle and the lead drill provides a clearance for the lathe center points. The axes of the two center holes must coincide with each other so that the lathe centers will seat firmly on the conical surfaces of the holes. There are several methods of drilling the center holes in the work. See Figure 6-3.

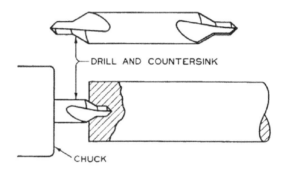

Figure 6-2 Here is an example of a combination drill and countersink, or center drill. It is used for centering parts for turning.

Figure 6-3 Here is a bench-top lathe with a tailstock drill chuck attached to the tailstock spindle.
Courtesy of Sherline Products

Types of Centers

The headstock center of the lathe always rotates with the lathe spindle; therefore it is called a live center. Tailstock centers may rotate with the workpiece and these are also referred to as live centers; however, frequently they are stationary or dead centers. See Figure 6-4.

Before inserting the center in place, the taper in the headstock must be thoroughly cleaned. The cleanliness of the taper as should be checked by using the bare hand or fingers. The presence of nicks or small particles of dirt or metal cannot be detected through a rag or cloth placed on the tapers; however, they can readily be felt by the fingers or the bare hand. All taper fits on any machine tool should be checked in this manner. If the taper is dirty, or badly nicked, it will not hold and it will not provide the desired accurate alignment of the mating parts.

Figure 6-4 Here is an example of a live center attached to the tailstock. The shape of this center provides enough clearance for the cutting tool.
Photo by James Harvey

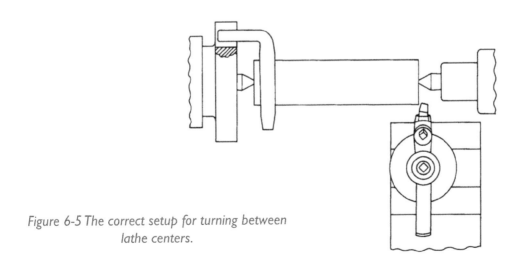

Figure 6-5 The correct setup for turning between lathe centers.

Setting Up the Work

The correct setup for turning between centers is shown in Figure 6-5. A driver plate is mounted on the spindle nose of the lathe. The driver plate is a small face plate that is used to drive the lathe dog.

Lathe dogs transmit the drive from the driver plate of the lathe to the workpiece when turning between centers. Several different types of lathe dogs used, are shown in Figure 6-6.

When a nonrotating, or dead, tailstock center is used, apply a lubricant. No lubricant is required if a live tailstock center is used nor is a lubricant used on the headstock center. Place the workpiece between the centers. Position the lathe dog with the bent end, or tail, between one of the slots on the faceplate and clamp it to the work. The tailstock spindle is adjusted until the work seats on both centers, without binding, and the spindle is clamped in place. It is important to check this adjustment frequently during the course of the turning operation. When a heavy cut is taken, the workpiece may absosrb some of the heat generated by the cutting action, which will cause it to expand in length enough to make the centers bind. When this occurs, the tailstock center will overheat and, consequently, be ruined. When a heavy roughing cut is followed by a light finishing cut, the work will lose heat and contract, causing the adjustment of the centers to be too loose.

A cutting tool is selected and clamped in the tool post. The overhang of the cutting edge beyond the tool post should be kept to a minimum in order to make the setup as rigid as possible. The nose of the cutting tool should be positioned "on center" or at the height of the lathe centers, as shown in Figure 6-7. The cutting tool is usually, but not always, held about square with the workpiece as in Figure 6-5. When turn-

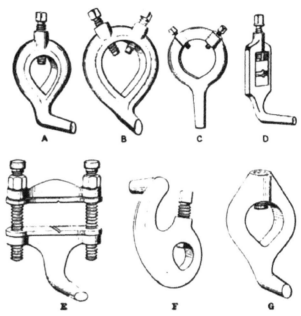

Figure 6-6 Types of dogs, or drivers, commonly used in connection with lathe work.

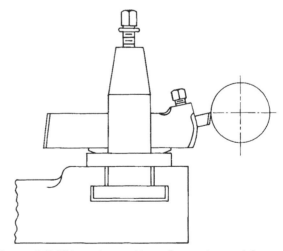

Figure 6-7 This is the position of a cutting tool for turning. The nose of the tool should be "on center" or at the height of the lathe centers.

ing close to the lathe dog or to a large shoulder or for other reasons, the cutting tool may be held at an angle.

Working with Chucks

There are many different designs and types of chucks used in a machine shop. For example, there are drill chucks for holding straight-shank twist drills, magnetic chucks, vacuum chucks, etc. The three-jaw universal chuck, four-jaw independent chuck, and collet chucks are commonly used on lathes to clamp workpieces and hold them in alignment when turning, facing, thread cutting, drilling, boring, reaming, and cut off. Short workpieces are held only by the chuck, while longer ones are given additional support by the tailstock center or by a steady rest.

The Three-Jaw Universal Chuck

Three-jaw chucks are similar to the chuck on a power hand drill. The operator uses a key to open and close the jaws. The jaws move in the slides that are machined onto the face of the chuck body. Each slide and jaw is numbered, for each jaw must be assembled into its corresponding slide. This is important because jaws on chucks are reversible. See Figure 6-8.

Fgure 6-8 A three-jaw chuck provides a quick way to setup work on a lathe.
Photo by Tom Lipton.

Because of the clearances required to allow this mechanism to operate without binding and because of wear, three-jaw chucks do not always hold the workpiece accurately in terms of a few thousandths of an inch. In many cases this is good enough, but if it is not, one solution is to use a four-jaw chuck.

The Four-Jaw Independent Chuck

The four-jaw independent chuck is a very versatile chuck with a wide range of applications on lathe work. Each jaw works independently by means of a separate screw turned by the chuck key. Parts can be held firmly and, if required, very accurately, in this chuck. The accuracy with which a part can be held is dependent entirely on the condition of the surface on the work. When the procedure is understood, four jaw chucks can usually be adjusted accurately within a fraction of a thousandth.

Mounting work in the four-jaw chuck. The chuck jaws should be opened just far enough to admit the workpiece by observing the position of each jaw in relation to concentric grooves that are machined into the face of the body. The jaws are then closed to hold the work lightly. Place the shank end of a tool holder, or a piece of metal that is resting in the tool post, against the work opposite one of the chuck jaws as shown at A, Figure 6-9. The tool holder must not be clamped tightly in the tool post as it must be able to be pushed aside with little effort. Then, turn the chuck manually. The high spot on the work pushes the shank of the tool away, resulting in an opening, D (shown in B, Figure 6-9). This occurs when the low spot is opposite the shank. Adjust the jaw with the largest opening between the shank and the work, and its opposing jaw, until the openings opposite these two jaws are equal. The other two jaws then are adjusted likewise. Finally, the fine adjustment is made again, first by adjusting two opposing jaws, then the other pair of opposing jaws. For greater precision, a dial indicator can be used in place of the tool holder to center the work.

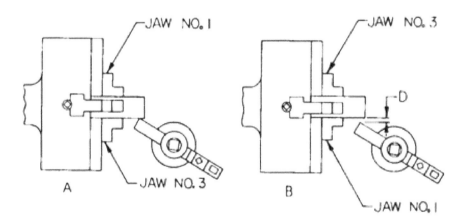

Figure 6-9 Here is the procedure for adjusting workpiece in a four-jaw independent chuck.

Reversible Chuck Jaws

Most chucks are reversible, able to hold both inside and outside diameters of workpieces. It is important to follow the manufacturer's suggestions when reversing the jaws on the chuck. See Figure 6-10.

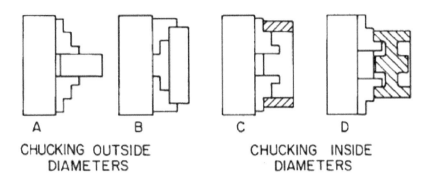

A   B   C   D

CHUCKING OUTSIDE   CHUCKING INSIDE
DIAMETERS   DIAMETERS

Figure 6-10 Schematic drawing showing the basic methods of holding the workpiece in both three-jaw independent chucks and four-jaw universal chucks. Only two chuck jaws are shown for illustration purposes. Any set of steps on the jaws can be used to hold parts.

Collet Chucks

Spring collet chucks are highly accurate chucks. They should be used to clamp on finished surfaces that are accurate to within a few thousandths of an inch of the size of the collet used. In other words, the workpiece has to fit the collet. Different size workpieces require different collets. The spring collets exert an even and uniform pressure around the workpiece, which minimizes the distortion of thin-walled parts. A special collet adapter is inserted in the spindle nose and the collet is inserted in the adapter. The drawbar is inserted in the rear of the spindle. When the drawbar is turned, it draws the collet into the spindle by the action of the threads. The taper on the adapter (corresponding to the taper on the collet) causes the collet to close as it is pulled into the spindle by the drawbar.

The diameter of the workpiece held in a spring type collet should be within approximately ± .005 in. (±0.13 mm) of the diameter for which the collet is made. Figure 6-11 illustrates what happens when the work piece is too large or too small. The collet will not grip the workpiece firmly, except at the ends of the collet holding surfaces which, as a result, may cause the machined surfaces to be inaccurate. The workpiece surface on which the collect grips should be a finished surface to obtain the accuracy and the maximum holding power for which the collect has been designed. See Figure 6-12.

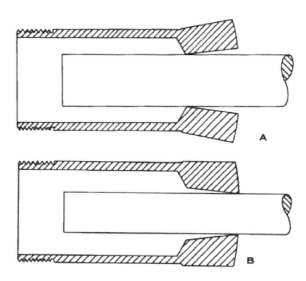

Figure 6-11 Effect of holding oversize and undersize workpieces in a spring collet.
A. Workpiece oversize; B. Workpiece undersize.

Figure 6-12 The box in the foreground holds a collection of collets. To keep the workpiece aligned
properly, choose the correct size collet.
Courtesy of Sherline Products

Turning

Two types of cuts are normally taken when turning: roughing cuts and finishing cuts. See Figure 6-13. In either case, the cutting tool is positioned to cut the required diameter by using the trial-cut procedure. Here is a description.

Figure 6-13
Turning operations consist of the roughing followed by a finishing cut to obtain the desired dimensions of the workpiece.

Adjust the cutting tool to take a shallow cut using the cross slide to position the tool. Take care to make sure that this cut will leave the work oversize. Make a ½-inch in length cut. With the lathe spindle stopped, measure the diameter of the work using micrometer or calipers. The carriage is then moved back toward the tailstock in order to position the tool for taking a second cut. Before taking this cut, adjust the cutting tool to cut the workpiece to the desired diameter by moving the cross slide the exact distance required. This distance is one-half the difference between the measured diameter of the trial cut and the desired diameter. The distance that the cross slide is moved can be accurately determined by reading the micrometer cross-feed dial. Some cross-feed dials are direct reading, while others must be moved one-half of the difference between the trialcut diameter and the desired diameter. After the cutting tool has been set to take the second cut, the longitudinal feed is engaged and another 1/2-inch-long cut is taken. The resulting diameter is measured. If the diameter of the work is to size, the longitudinal feed is engaged and the cut is taken to the required length. It is sometimes necessary to take several trial cuts before the desired diameter on the work is obtained.

The entire workpiece should be "roughed out" before taking any finish cuts by taking one or more roughing cuts on each diameter that is to be turned. The roughing cuts should leave the diameter of the workpiece oversize a predetermined amount, which in general practice, may vary from approximately .020 to .060 inch. Shoulder lengths should be left short, or "undersize," in order to provide metal for the finish-turning operation. This amount is usually, approximately, .015 to .030 inch.

Finished Cuts. The finishing cut is often taken with the same cutting tool as was used to take the roughing cut. This procedure requires fewer changes of cutting tools. It is sometimes best, however, to use a separate tool for finish turning since less tool wear then occurs.

The cutting speed for finish turning can usually be faster than for rough turning because of the decrease in the depth of cut; however, not all lathes can be adjusted to obtain this speed differential. Carbide cutting tools should be operated at a cutting speed that is fast enough to produce a good surface finish. When these cutting tools are operated above a certain critical cutting-speed range, an excellent surface finish can easily be obtained. This critical cutting-speed range is dependent on the grade of carbide used and the nature of the work material. In general, it is in the range of 150 to 200 fpm. When slower cutting speeds must be used, the surface finish obtained can be improved by using a good cutting fluid and by using a cutting edge that is in good condition. This usually means that the cutting edge must be replaced before excessive wear has taken place.

Shoulder Turning

There are three steps in shoulder turning:
- Locating the shoulder on the work
- Turning the smaller diameter to size
- Facing the shoulder to length

The length of the shoulder can be laid out using a hermaphrodite caliper as shown in Figure 6-14. A scratch, or mark, made with the point of the cutting tool can also be used to mark the location of the shoulder. Often, the location of the shoulder is measured as the cut progresses, using a steel rule or a hermaphrodite caliper. A groove cut with a necking tool, A, Figure 6-15, can serve to locate a shoulder.

Several methods of turning shoulders are illustrated in Figure 6-15. At A, a groove is first cut with a necking tool in order to establish the length of the shoulder. The small diameter is then cut to size by one or more cuts with the turning tool. When this method is used, all of the required shoulders are "blocked out," and the adjacent cylindrical surfaces are cut to size. On slender workpieces, a very slow feed is necessary to cut the grooves in order to avoid chatter. (continued on the next page)

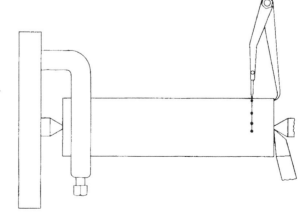

Figure 6-14
A hermaphrodite caliper is used to mark a workpiece for turning.

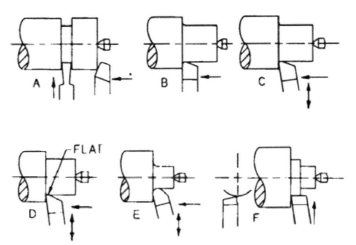

Figure 6-15. Procedures for turning shoulders.

A faster and even simpler method of cutting a shoulder is shown at B. The cutting tool is positioned to have a zero-degree lead angle. The shoulder is formed by the cutting edge when turning the smaller diameter. Shoulders turned in this manner are not necessarily perfectly square. When a shoulder must be square, the method illustrated at C is recommended. Usually the shoulder is first roughed out using either of the two previous procedures. The tool is then reset with the nose leading, as shown at C. It is set to cut the small diameter to size, using the trial-cut procedure. The cross feed dial reading used to take this cut, is noted. When the tool approaches the shoulder, feed the nose of the tool by hand, longitudinally, into the shoulder. When it has penetrated the shoulder enough to take a light cut, it is fed out using the cross slide. The length of the shoulder is then measured, and the tool is moved longitudinally, just enough to remove the excess metal from the shoulder. It is then fed into the work with the cross slide. When it approaches the corner of the shoulder it is carefully fed by hand, until the reading of the cross-feed dial is the same as was previously noted. In this manner, the tool will not undercut the corner and a smooth blend can be obtained between the shoulder and the previously turned diameter. It is sometimes an advantage to turn the compound rest 90 degrees for this operation. In this position, the compound-rest slide will move parallel to the longitudinal feed, and it can be used to set the depth of cut for the final finishing cut on the shoulder. Very accurate shoulder lengths can be obtained using this method of setting the tool. This same procedure can be used to cut a shoulder having a sharp corner, as shown at D, or to cut a radius in the corner, as shown at E. The tool, at D, must be ground with a flat on the end-cutting edge in order to finish turn the small diameter, otherwise it will cut threadlike grooves on this surface. Very large radii are cut using a large forming tool that is similar to the tool at E. The corner is first blocked out using a turning tool, such as shown at B. It is sometimes faster to block out the corner with a facing tool, as shown at F.

Steady Rests

One of the most useful lathe accessories is the steady rest. The purpose of the steady rest is to support long, slender parts, in a lathe. The additional support reduces the deflection of the workpiece caused by cutting forces, improving accuracy and reducing chatter. On very long and slender workpieces a steady rest is sometimes required just to keep the workpiece from sagging under its own weight. The steady rest may be used to support workpieces that are held between centers, with one end held in a chuck and the other on the tailstock center, see Figure 6-16. In order to leave the end face free for machining, one end is held in a chuck while the other end is supported only by the steady rest.

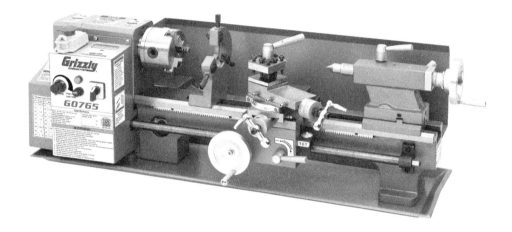

Figure 6-16 A steady rest provides support for long workpieces.
On this lathe, the steady rest is between the headstock and the tool holder.
Courtesy of Grizzly Industrial, Inc.

Other Lathe Functions

Drilling

The first step in drilling a hole on a lathe is to start the hole with a combination drill and countersink, or center drill, see Figure 6-2. The center drill is held in a drill chuck which is mounted in the tailstock spindle. The work is rotated by the headstock spindle and a high spindle speed should be used to drill this starting hole. The hole is drilled by clamping the tailstock to the bed of the lathe and feeding the drill into the work manually by turning the tailstock handwheel. The center drill, being short and inflexible, will not deflect; but will start the hole true. The twist drill will be guided by the hole made by the center drill and, therefore, will also start true and without eccentricity.

After the starting hole has been made with the center drill, the hole is then drilled to the required size and depth with a twist drill. The spindle speed used to drill the hole should be appropriate both to the size of the drill and the material being cut.

Reaming

Reamers are used to obtain an accurate diameter as well as a good surface finish in holes that are being machined in a lathe. Usually the hole is bored prior to reaming in order to make certain that the resulting hole will be concentric with the rotation of the work. The reamer will follow the existing hole through which it cuts. The reamer is a finishing tool and is designed to remove only a small amount of metal. Approximately 0.008 to 0.035 inch should be left on the diameter of the hole for reaming. In other words, the hole should be 0.008 to 0.035 inch smaller in diameter than the size of the reamed hole. On engine lathes the reamer is fed manually through the hole by using the tailstock spindle. Cast iron is usually reamed dry; however, a good grade of cutting oil should be used for reaming most other metals including all of the steels.

Boring

The objective of boring a hole in a lathe is:

- To enlarge the hole
- To machine the hole to the desired diameter
- To accurately locate the position of the hole
- To obtain a smooth surface finish in the hole

The boring tool is held in a boring bar which is fed through the hole by the carriage. A typical boring tool is shown in Figure 6-17. Variations of this design are used, depending on the job to be done. The lead angle used, if any, should always be small. Also, the nose radius of the boring tool must not be too large. The cutting speed used for boring can be equal to the speed for turning. However, when the spindle speed of the lathe is calculated, the finished, or largest, bore diameter should he used. The feed rate for boring is usually somewhat less than for turning to compensate for the lack of rigidity of the boring.

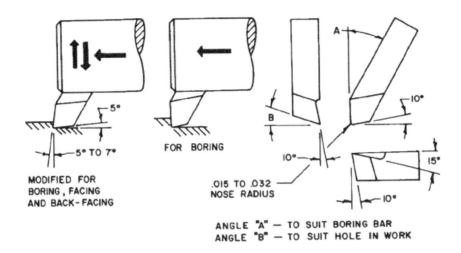

Figure 6-17 Design of a typical high-speed steel boring tool. The design is typical.

The boring operation is generally performed in two steps; namely, rough boring and finish boring. See Figure 6-18. The objective of the rough-boring operation is to remove the excess metal rapidly and efficiently, and the objective of the finish-boring operation is to obtain the desired size, surface finish, and location of the hole. The diameter of the hole can be measured with inside calipers and outside micrometers, or inside micrometers can be used to measure the diameter directly.

Figure 6-18. Here's an example of a boring tool cutting tapers in a workpiece.

Cutting Off

The objective of the cut-off operation is to sever the end of the workpiece which is protruding from the chuck. A cut-off tool, Figure 6-19, performs the cut-off operation. The end cutting edge is ground at a 30 to 15° angle in order to cut off the part without a burr. The burr remains on the stock that is held in the chuck and can be removed by feeding the tool a short distance beyond the point where the stock is separated. The width of the cut-off blade is dependent upon the size of the cut-off tool. It should be kept as small as possible. Although both larger and smaller widths are used, the average width of the cut-off blade is from 1/8 to 1/4 inch. The blade has a back taper of one degree on each side in order to provide clearance for the tool with respect to the sides of the groove made by this tool. The face of the tool is made flat, although a very large radius (equal to the radius of the grinding wheel used to prepare this surface) is sometimes ground on the face. This provides a very small back-rake angle.

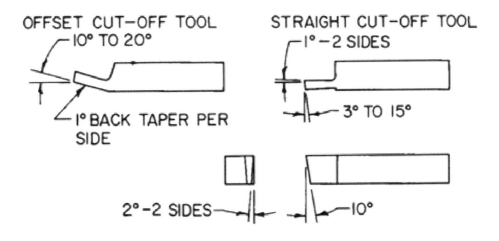

Figure 6-19 Design of a cut-off tool.

Cutting Speeds. The cutting speed used to cut off stock can usually be as fast as recommended for turning. Occasionally, a very deep cut must be taken to cut off a part, in which case the cutting speed should be reduced. This is done more in the nature of a safety measure than to decrease the cutting load on the tool. The cutting load, or cutting force, on the tool will not be reduced by decreasing the cutting speed, but by reducing the feed. For this reason, a relatively light feed should always be used to perform the cut-off operation. For very small cut-off tools, a feed rate as low as .0005 inch per revolution is sometimes used. Large cut-off tools can operate satisfactorily using a feed of .006 to .010 inch per revolution. For average conditions a feed rate of .001 to .005 inch per revolution is recommended.

The correct setup for an offset cut-off tool is shown in Figure 6-20. For both offset and straight cut-off tools, the blade must always be positioned perpendicular to the workpiece axis and the end cutting edge must be on center.

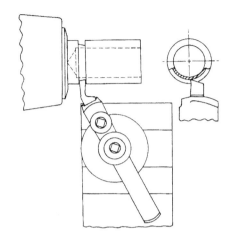

Figure 6-20. This is the correct setup for an offset cut-off tool. The blade must be perpendicular to the work and on center.

Cutting Threads

To cut threads using a lathe, you will need to manipulate the gears on the machine to control the work. Some lathes come with quick-change gear boxes while smaller lathes usually have a gear system that must be installed for thread cutting. The lathe manufacturer will provide instructions for setting up the gears and making the cuts. See the sidebar for standard thread definition that will be useful when cutting threads.

Screw-Thread Terms

The definitions of the principal terms relating to screw threads listed below (see Fig. 6-21):

Major Diameter—The largest diameter of a straight screw thread. The term major diameter applies to both internal and external threads.

Minor Diameter—The smallest diameter of a straight screw thread. The term minor diameter applies to both external and internal threads.

Pitch—The distance from a point on a screw-thread profile to a corresponding point on the next thread profile, measured parallel to the axis. The pitch is equal to one divided by the number of threads per inch.

Lead—The distance that a screw thread advances axially in one turn. On a single thread screw the pitch and the lead are equal; on a double thread screw the lead is twice the pitch; on a triple thread screw the lead is three times the pitch, etc.

(continued on the next page.)

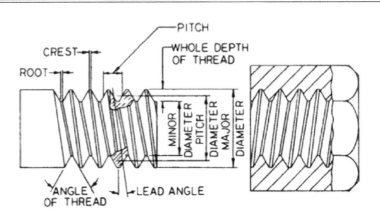

Figure 6-21 Illustration showing screw-thread terms.

Angle of Thread—The angle included between the sides of the thread measured in an axial.

Lead Angle—This is the angle made by the helix of the thread at the pitch diameter, measured in an axial plane.

Crest—The top surface joining the two sides, or flanks, of the thread.

Root—The bottom surface joining the two adjacent sides or flanks of the thread.

Right-Hand Thread—A thread with the grooves cut so that a machine screw or a bolt would have to be turned in a clockwise direction, when viewed from the head end, in order to assemble into a threaded hole.

Left-Hand Thread—A thread with the grooves cut so that a machine screw or a bolt would have to be turned in a counterclockwise direction, when viewed from the head end, in order to assemble into a threaded hole.

Single Point Cutting Tool. The start of a good thread is the thread-cutting tool. It must have the correct shape, or profile, in order to cut an accurate thread. Usually this shape corresponds exactly with the shape of the thread to be cut. Accurate thread profiles can be ground on tool and cutter-grinding machines. Carbide thread-cutting tools are usually purchased already formed to the required shape. High-speed steel thread-cutting tools are usually ground to the required shape. Whether the thread-cutting tool is ground to the required shape on a cutter and tool-grinding machine, or by hand, on a pedestal grinder, it should be carefully inspected before it is used.

149

The design of a high-speed-steel cutting tool for cutting American Standard Unified and National Thread Forms is shown in Figure 6-22. This tool has a 60-degree included angle and the width of the flat, F, is made appropriate to the pitch of the thread to be cut. View A, is a top view of the tool showing the shape, or profile. An alternate shape is also shown. View B, is a side view

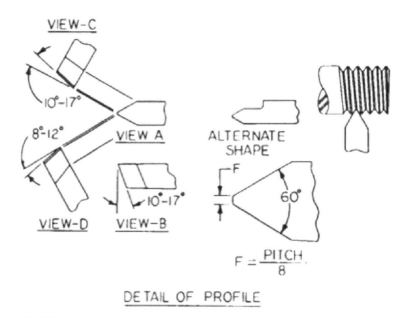

$$F = \frac{PITCH}{8}$$

DETAIL OF PROFILE

Figure 6-22 Thread-cutting tool for American Standard Unified and National Threads.

Figure 6-23 A cutting tool finishing cutting threads.

of this tool, showing the angle made by the flat. This angle should be 10 to 17 degrees depending upon the hardness of the metal being cut. The right side of the tool is shown at C. View D is a true auxiliary view obtained when sighting parallel to the right side-cutting edge of the threading tool. It shows the right side-cutting edge relief angle which should be 8 to 12 degrees when cutting right-hand screw threads. The face of the thread cutting tool is ground fiat, or without a rake angle. It should, however, be ground smooth in order for all the cutting edges to be sharp. See Figure 6-23.

Cutting the Threads. Referring to Figure 6-24:

1. Start the lathe spindle. With the thread cutting tool just left of Position 1, move the cross slide inward until the tool just touches the workpiece. This position of the cross slide is the zero reference position. Before each cut through the thread groove, the cross slide must be at the zero reference position.

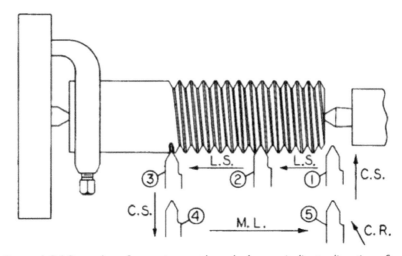

Figure 6-24 Procedure for cutting an threads. Arrows indicate direction of motion:
L.S. Feed by lead screw; C.S. Feed by cross slide; M.L. Manual longitudinal feed;
C.R. Compound-rest slide feed.

2. Move the thread cutting tool to Position 1 to be clear of the end of the workpiece. Establish the depth of cut by moving the thread cutting tool inward the required distance with the compound rest. To obtain a precise depth of cut setting, the compound rest micrometer dial should be used. For the first scratch cut, the depth of cut should be .001 inch to .003 inch (0.03 mm to 0.08 mm).

3. The carriage advances the tool into the work to cut the thread groove, as shown by Position 2. Be prepared to withdraw the tool from the thread groove and to disengage the split nut at the end of the cut.

4. At the end of the cut, Position 3, quickly turn the cross feed crank handle to withdraw the thread cutting tool with the cross slide. When the tool is just clear of the thread groove, stop the carriage movement. This order can be reversed when the thread ends in a groove or neck cut into the workpiece to provide clearance for the tool; i.e., first stop the carriage and then withdraw the tool. When this step has been completed, the tool should be in Position 4, clear of the workpiece.

5. Move the carriage manually from Position 4 to Position 5. When the tool is in Position 5, stop the lathe spindle and check to see if the required threads per inch are being cut. Place a rule against the thread grooves and count the number of grooves cut in one inch. If available, a thread pitch gage may be used. If necessary, correct the quick-change gearbox setting. This step is performed only on the first cut; it is not repeated on the following cuts.

6. At Position 5, set the depth of cut for the next cut by moving the thread cutting tool inward with the compound rest slide.

7. Move the cross side inward to the zero reference position (Position 1). To eliminate the effect of the clearance between the cross feed screw and the nut, called backlash, the movement to the zero reference position must be made with the cross slide moving inward toward the lathe axis, or toward the workpiece.

8. The thread cutting tool is now prepared to take the next cut. The lathe spindle needs to be stopped only to measure or gage the size of the thread.

The size of the thread is not usually measured directly when it is in the process of being turned in a lathe. The procedure of bringing it to size is a cut-and-try procedure. The appearance of the thread will give an indication when it is getting close to the finish size. When it is judged to be close to the finish size, it is gaged with the ring gage, snap gage, or mating part, after each cut, until the gage fits properly. Essentially the same procedure is used when measuring the threads by the three-wire method or with thread micrometers. See Figure 6-25

Figure 6-25 A triangular file cleans up a damaged thread.

Tip from a Pro

Turning Between Centers

Text and photos reprinted by permission of
J. Randolph Bulgin
and *The Home Shop Machinist*
(May/June 2009).

LBS (Long Boring Story) coming up so bear with me. I was a Machinery Repairman Third Class Petty Officer in the Navy when my ship went into the Portsmouth Naval Shipyard for an overhaul. The crew lived aboard during the six week process and I, along with a couple of ship fitters, found that the shipyard was so lacking in workers they were willing to hire us on an hourly basis when we were off duty. It was a good deal for us. We made more wages than any of us had ever heard of; I think we were paid something on the order of three bucks an hour (this was in 1960) and it was a great place to escape from the constant chipping and welding that was going on in our living quarters. The arrangement didn't last long but that is another story.

On my first day as a hired machinist in a real machine shop, I was assigned the job of making some sheave pins for crane pulleys. I was told where to go to get the material and where the lathe, a 20" American Pacemaker, was located. I went happily off to get the 3" diameter piece of 4140 steel, expecting to be given a piece long enough to chuck one end while machining the required diameters on the part. The stockroom attendant sawed my material to exactly the length of the finished part! There was just enough extra length of the material to take light facing cuts on either end. How am I going to hold on to this thing to machine the full length? The answer was, of course, to set the part up so that I could turn it between centers. I have turned many parts since then using this method, and it is considered, at least by me, to be one of the basic processes for a machinist to learn. Here is one way of doing it.

First, cut your material to length using the same stingy process the stockroom attendant at the shipyard in my story did. Saw to length plus about 1/16". If you are sure your saw will produce a square cut, you can leave a little less. All that is required is that you have enough material to face the ends clean and stay within your length tolerance. Next, machine center holes in both ends. **Photo 1** is a picture of the combined drill and countersink used to produce the center hole. It is important that the center hole have the following

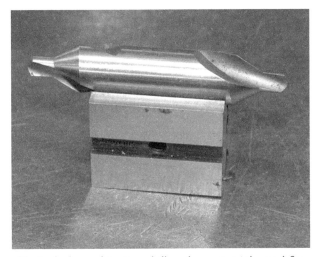

Photo 1 *A combination drill and countersink used for machining centers in a part.*

characteristics:

- Be in the center of the part
- Deep enough to ensure it won't come out when cutting force is applied
- Have a smaller diameter hole in the center so that the point of your lathe center will not be in contact with the bottom of the hole
- Clean and free of chips and debris that might cause the part to run out

Photo 2 shows the center hole as drilled in one end of the material. If the job is too long to support securely in a chuck and if your lathe does not have a hole through the spindle large enough to accommodate it, you have a couple of choices. You can either use a steady rest to support the work while drilling the centers, or you can locate and center punch the locations and drill them with a hand drill. I have done both, and both methods work. Facing the part gives you a nice finished end and it also provides a clean and square surface in which to drill your center hole.

Photo 2 Drilling the center hole in one end of the material.

Photo 3 A center for the headstock. This must be re-machined every time it is removed and replaced in the chuck.

After the workpiece is prepared you must prepare the machine. When this is done you will have

- a live center in the headstock that will accept the center hole you have in the part and is running dead true with the axis of the lathe;
- some means of driving the part such as a face plate, a drive plate, or a chuck with jaws that will engage and drive the lathe dog;
- a tail center that will accept the center hole in the other end of the part and is in perfect alignment with the center in the headstock. This may be either a ball bearing live center or a hardened center with adequate lubrication.

One method of preparing the headstock center is shown here. First, make sure you are finished with the process of preparing the material. Once you machine the center in the headstock, you should not remove it until the job is done. If you do have to remove it for any reason, you must re-machine the center when you replace it. The accuracy of your job will depend to a great deal upon the accuracy of these centers. When you are ready for the center, mount an appropriate piece of material *(Photo 3)* in either a three- or four-jaw chuck. Note that the material has a shoulder that bears against the front of the chuck jaws. This will ensure that the center will not be forced back into the chuck under the pressure of the cut. Set the compound at 30° from the axis of the machine *(Photo 4)*, and machine the point of the center as shown in ***Photo 5***. Many lathes, partic-

Photo 4 Setup for machining a 60-degree center.

Photo 5 Machining the center. The tool must be hand fed using the compound rest.

155

ularly the smaller bench or tool room lathes, are equipped with a center that is inserted into the headstock of the machine. This works well, but I prefer the method shown here. With this method, you can control the length the center protrudes from the chuck, which is sometimes an advantage. You can also be sure that the center is running in perfect concentricity with the axis of the lathe, and you can use one of the chuck jaws to drive the lathe dog. Whatever works for you is the method to use.

Now mount the lathe dog on the work. **Photo 6** is a picture of a small collection of dogs I keep handy. Place the part between the centers as shown in **Photo 7**. Here, I am using

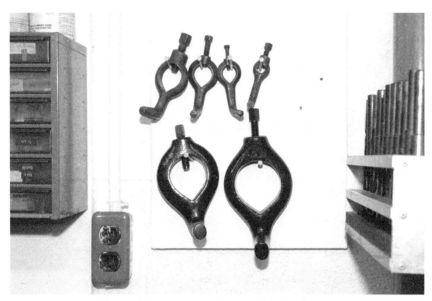

Photo 6 A sampling of bent tail lathe dogs.

Photo 7 A part set up between centers.

a ball bearing live center in the tailstock, which, in my opinion, is the best choice for this type of work. Dead centers are typically more accurate and are widely used in grinding operations, but the loss in accuracy in lathe work is minimal and is, again in my opinion, more than offset by the convenience of not having to keep the centers lubed.

Now on to the last preparatory operation of setting up work to turn between centers, which is ensuring that the job is running true and without taper. If the tail center is offset at all, you will machine a tapered part. In fact, this is a good and accepted method for producing tapers but that is a subject for another time. You want the part we are machining to be straight, and here is how you can make it straight. After you have checked all of the variables:

- The dog screw is tight. (Watch out for that dog. It will bite if you get your elbow into it!)
- The dog tail is firmly engaged against the chuck jaw or drive plate slot.
- The tail center is tightened so that there is no slack in the setup.
- You have checked that the compound will not try to occupy the same space as the lathe dog.

You are ready to take your first cut!

Begin by machining a short diameter at the tail end of the work. Set the cross-slide or the DRO to zero and back out. Then move to the other end as shown in *Photo 8* and machine another diameter at the same cross-slide setting. Measure the two diameters *(Photo 9)* and note the difference. If there is no difference, or if the difference is within the tolerances you are given for taper, proceed with the job. If there is a difference, you must make some adjustments.

For the sake of this instruction let us say that the diameter of the work at the tail end is .004" larger than the diameter at the other end. You must move the tail end .002" closer to the cutting

Photo 8 Test cuts to determine the amount of taper — if any.

Photo 9 Measuring for taper.

Photo 10
Adjusting to remove taper.

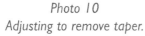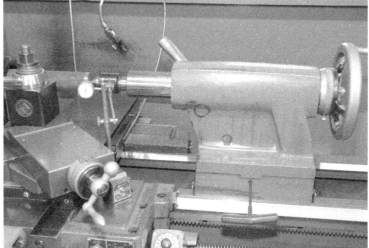

tool in order to make the two ends the same. Set up a dial indicator *(Photo 10)* so it bears against the machined diameter on the tail end. You must then loosen the clamp that clamps the tailstock to the lathe bed and, by tightening and loosening the screws at the base of the tailstock, move the workpiece the proper amount in the proper direction. The red-handled hex wrench in the photo is what adjusts tailstock set-over on my lathe. Yours may be different, but there will be some method provided for this adjustment to be made. After you get the desired reading on the dial, tighten everything back down again, and repeat the process by taking another light cut on the diameters at each end. Make any further necessary adjustments and you are all set.

Most any machining process, particularly shaft work, which can be done in a chuck, may be done between centers. But as in everything you do in the machine shop, pay attention to what you are doing. For example, if you are threading and you want to remove the part from the lathe to check a fit, be sure you put it back with the dog having the same orientation to the driving surface. And this bears repeating. That lathe dog will bite you if you let it!

7   Milling Machines

The milling process is used to produce a variety of surfaces by using a circular-type cutter with multiple teeth or cutting edges. Unlike lathes, which turn the workpiece while keeping the cutting tool stationary, the motor on a mill turns the cutting tool while the workpiece remains stationary.

Milling machines, as a class of machine tools, are very versatile. They are capable of machining one or two piece lots as well as parts on a large-volume production basis. The inherent advantage of the milling process is the circular cutter, which is economical and has a high metal removal rate since it can bring a large number of cutting edges into the cut in a relatively short space of time. While each manufacturer's product is slightly different from the others, mills can be divided into two broad categories: *horizontal mills* and *vertical mills*. The terms horizontal and vertical refer to the way the cutting tool is mounted on the machine. Horizontal mills are usually found in large production shops. For the home hobbyist, vertical mills make the most sense because they are available in a number of sizes, including bench-top models, and they can perform the milling functions most hobbyist need. The rest of this chapter will cover the description and operation of vertical type mills.

Principal Parts

The parts of a typical mill are shown in Figure 7-1. Although products vary from manufacturer to manufacturer most consist of a central column that holds the motor and cutting tool at the top and a table that holds the work at the bottom. Tables have the ability to move from left to right as well as toward and away from the column. These are often referred to as the x axis (side to side) and the y axis (front to back) of the machine. In addition, on bench-top models the motor and cutting tool moves up and down—the z axis. This range of motions helps when it is time to align a workpiece with the cutting tool. When purchasing a mill, make sure the one you want has sufficient clearance between the table and the cutting tool to accommodate the work you want to perform.

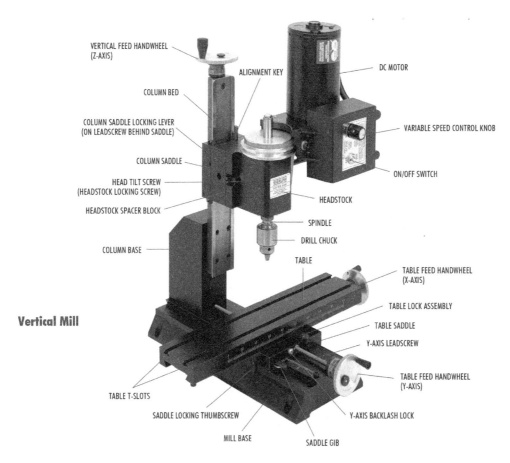

VERTICAL FEED HANDWHEEL (Z-AXIS)

ALIGNMENT KEY

DC MOTOR

COLUMN BED

VARIABLE SPEED CONTROL KNOB

COLUMN SADDLE LOCKING LEVER (ON LEADSCREW BEHIND SADDLE)

COLUMN SADDLE

ON/OFF SWITCH

HEAD TILT SCREW (HEADSTOCK LOCKING SCREW)

HEADSTOCK

HEADSTOCK SPACER BLOCK

SPINDLE

COLUMN BASE

DRILL CHUCK

TABLE

Vertical Mill

TABLE FEED HANDWHEEL (X-AXIS)

TABLE LOCK ASSEMBLY

TABLE SADDLE

Y-AXIS LEADSCREW

TABLE FEED HANDWHEEL (Y-AXIS)

TABLE T-SLOTS

SADDLE LOCKING THUMBSCREW

Y-AXIS BACKLASH LOCK

MILL BASE

SADDLE GIB

Figure 7-1 This is a popular bench-top mill for home hobbyists.
Courtesy of Sherline Products Inc.

Cutter Names

   Milling cutters can roughly be classified as arbor-mounted cutters—usually found on horizontal mills— end-milling cutters, and face-milling cutters. There are various standard types, styles, and sizes within each classification from which a selection can be made to suit most applications. The nomenclature can become confusing because milling cutters are also called mills and an end milling cutter is often called an *end* mill. The word, "mill" is also used, of course, to designate the actual milling operation—for example, "to mill a surface."

Milling Cutter Materials

The cutting edges of milling cutters are primarily made of high-speed steel or from carbides. Many milling cutters are made from solid high-speed steel, including most general-purpose end mills and arbor-mounted cutters. End milling cutters and arbor mounted cutters made for high production often have carbide cutting edges. Face milling cutters also may have high-speed steel or carbide cutter blades, or carbide indexable inserts.

High-Speed Steel. High-speed steels are a group of highly alloyed tool steels characterized by their ability to retain a high level of hardness and wear-resistance at temperatures up to approximately 1,100° F (590° C), where other tool steels will soften and fail. When annealed, high-speed steel can be machined into the shape of the cutting tool and then hardened again. High-speed steels are very deep-hardening, allowing them to be sharpened many times without a significant loss in hardness.

Carbide Cutters. Carbides are harder than high-speed steel and can retain their hardness at a higher temperature. As a result, much faster cutting speeds can be used when milling with carbides. Carbides, however, are more brittle and less shock-resistant. For this reason, greater care must be used with carbide milling cutters on the job.

End Milling Cutters

Although there are a variety of cutters available for mills, including circular cutters and slitting saws shown in Figure 7-2, end milling cutters constitute a large group of milling cutters made to a variety of shapes and sizes. They are characterized by having cutting edges on the end face as well as on the periphery. Also, they are always held in the milling-machine spindle by a collet chuck or some kind of adaptor. Among the most versatile of cutting tools, end mills are used to mill plane surfaces, slots, profiles, and three-dimensional contours.

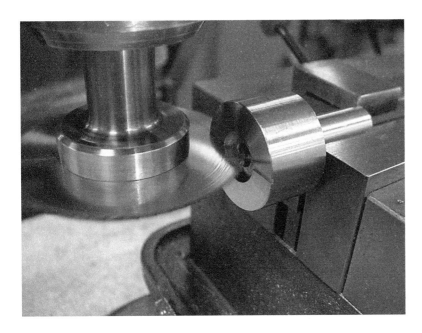

Figure 7-2
Here is an example of a slitting saw make a cut on the face of a part.
Photo by James Harvey

The elements of an end milling cutter are shown in Figure 7-3. The radial rake angle is generally small, as it is limited by the necessity of keeping the end cutting edges approximately radial. The helical rake angle and the helix angle are for practical purposes the same angle.

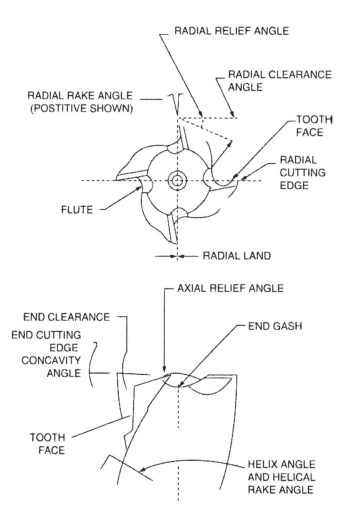

Figure 7-3
Elements of
end milling cutters.

The relief angle of the peripheral teeth, or the "radial relief angle," is determined by the size of the end mill and by the material to be cut. For most applications, including most steels and cast irons, the following radial relief angles (given in terms of the cutter diameter first followed by the radial relief angle) are recommended:

1/16 inch—22 degrees, 1/8 inch—17 degrees, 3/32 inch—14 degrees, 3/16 inch—14 degrees, 1/4 inch—12 degrees, 3/8 inch—12 degrees, 1/2 inch—11 degrees

Right- and Left-Handed Cuts

The flutes in an end mill may have a right- or left-hand cut as shown in Figure 7-4. The cut refers to the side of the flute on which the face of the teeth are located. Furthermore, end mills with a right-hand cut are designed to cut while rotating counterclockwise when viewed from the end having the teeth. Left-hand-cut end mills are designed to cut while rotating clockwise when viewed from this end. Although the helix may be right or left-handed for either cut, usually a right-hand-cut flute has a right-hand helix and a left-hand-cut flute has a left-hand helix. In this way the end-cutting edges have a positive rake angle.

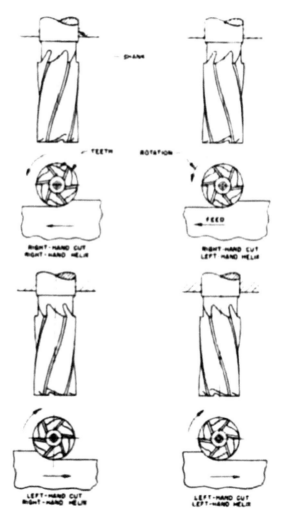

Figure 7-4 The four combinations of hand of helix and hand of cut for end mills.
Courtesy of Cincinnati Milacron

Number of Flutes. Most standard end milling cutters are made to have two or four flutes in sizes up to approximately 1 inch. (See Figure 7-5.) Larger end mills, up to 2 inches in diameter, are made with six or eight flutes. Three-fluted, center-cut-type end mills are made with diameters up to 3 inches. Increasing the number of flutes on the end mill helps to stabilize the cutter when milling slots and allows a faster feed (inches per minute) to be used. The flutes must, however, be large enough to provide adequate space for the chips.

The most common center-cutting-type end mill is the *two-fluted end mill,* which in some shops is called a *two-lipped end mill.* The conventional square end two-fluted end mill is the least expensive and the most easily sharpened of the center-cutting-type end mill. As it can easily be sunk directly in to the work piece like a drill, it is frequently used to cut keyseats and other slots that do not extend to an open end or shoulder. Two-fluted ball end mills are used to cut complex three-dimensional contours such as are encountered in dies and molds.

Figure 7-5 Shown is a set of end mills designed for a small bench-top mill.
Courtesy of Sherline Products, Inc.

Fly Cutters

Another type of cutter, called the fly cutter is shown in Figure 7-6. It offers many advantages in performing face milling operations for which it is intended. The cutting tool is a standard single-point, high-speed-steel tool, which can he sharpened by hand. When used with a fine table feed rate, this cutter will produce an excellent surface finish on most materials. Since only a single cutting edge engages the workpiece, the cutting force is light, enabling frail parts to be milled and setups to he used which are somewhat less secure than required when milling with a face milling cutter. On light, low powered milling machines, relatively large surfaces can be milled in a single pass that would otherwise require a series of passes with an end milling cutter.

Figure 7-6 A fly cutter, such as the one shown, is capable of producing a smooth finish.

Courtesy of Sherline Products, Inc.

Cutting Speeds for Milling

The cutting speed for milling is the speed at the periphery of the cutter as it is rotating. In the inch system of units, the cutting speed is given in terms of feet per minute, or fpm, which is sometimes called surface feet per minute, or sfpm. In the metric system, the cutting speed is in meters per minute, or m/min. The recommended cutting speeds feet per minute are given in Tables 7-1 through 7-4 for milling nitrous materials with high-speed steel cutters. Again, the recommendations in the tables are starting points, and may prove too aggressive for hobby-sized equipment. But you can use the information in the tables to calculate the cutting speed of the machine spindle in rpm. See "Calculating the Cutting Speed," 173 Experience using your equipment and working with specific materials will help you determine the correct cutting speeds. To obtain meters per minute, multiply feet per minute by .3048. For each material listed in the following tables, a range of values is given to account for the shop variables encountered.

Factors Affecting Cutting Speeds

In addition to the cutting tool material, the cutting speed depends primarily on the work material and its hardness. In general, an increase in the hardness of a material reduces the speed at which it can be cut. Since the hardness of a material is not always known in the shop, the material condition that is associated with a corresponding hardness in the table is given. The cutting speed is also influenced by the feed rate, and to a lesser extent by the depth of cut. Heavier cuts using a heavy feed require a slower cutting speed than do lighter cuts. Since the cost of replacing and sharpening a milling cutter is more than the cost of a single-point cutting tool, a longer tool life is more desirable for milling than for turning; therefore, the cutting speed for milling should he somewhat slower than for turning under the same tool and work material conditions.

When using carbide milling cutters, the grade of carbide used has an influence on the cutting speed that can be used. The correct grade, as recommended by the carbide producer or cutter manufacturer, must be used. Where they can be used, coated carbides can often cut successfully using a cutting speed that is 20 to 40 percent and sometimes up to 50 percent higher than the values given in cutting speed tables. In general, carbide cutters having indexable inserts are operated at a somewhat faster cutting speed than those having brazed-on carbide tips, or blades to which the carbide is brazed. Other factors to consider in selecting the cutting speed are the design of the milling cutter and the rigidity of the workpiece, the setup, and the machine. When starting out to mill a new material, start at the lower end of the range given in the table; then, as experience is gained, increase the cutting speed if necessary.

Table 7-1. Recommended Cutting Speeds in Feet per Minute for Milling Plain Carbon and Alloy Steels. See Milling Column.

Material AISI and SAE Steels	Hardness HB[a]	Material Condition	Cutting Speed, fpm HSS			
			Turning	Milling	Drilling	Reaming
Free Machining Plain Carbon Steels (Resulfurized)						
1212, 1213, 1215	100-150	HR, A	150	140	120	80
	150-200	CD	160	130	125	80
1108, 1109, 1115, 1117, 1118, 1120, 1126, 1211	100-150	HR, A	130	130	110	75
	150-200	CD	120	115	120	80
1132, 1137, 1139, 1140, 1144, 1146, 1151	175-225	HR, A, N, CD	120	115	100	65
	275-325	Q and T	75	70	70	45
	325-375	Q and T	50	45	45	30
	375-425	Q and T	40	35	35	20
Free Machining Plain Carbon Steels (Leaded)						
11L17, 11L18, 12L13, 12L14	100-150	HR, A, N, CD	140	140	130	85
	150-200	HR, A, N, CD	145	130	120	80
	200-250	N, CD	110	110	90	60
Plain Carbon Steels						
1006, 1008, 1009, 1010, 1012, 1015, 1016, 1017, 1018, 1019, 1020, 1021, 1022, 1023, 1024, 1025, 1026, 1513, 1514	100-125	HR, A, N, CD	120	110	100	65
	125-175	HR, A, N, CD	110	110	90	60
	175-225	HR, N, CD	90	90	70	45
	225-275	CD	70	65	60	40
1027, 1030, 1033, 1035, 1036, 1037, 1038, 1039, 1040, 1041, 1042, 1043, 1045, 1046, 1048, 1049, 1050, 1052, 1152, 1524, 1526, 1527, 1541	125-175	HR, A, N, CD	100	100	90	60
	175-225	HR, A, N, CD	85	85	75	50
	225-275	N, CD, Q and T	70	70	60	40
	275-325	Q and T	60	55	50	30
	325-375	Q and T	40	35	35	20
	375-425	Q and T	30	25	25	15
1055, 1060, 1064, 1065, 1070, 1074, 1078, 1080, 1084, 1086, 1090, 1095, 1548, 1551, 1552, 1561, 1566	125-175	HR, A, N, CD	100	90	85	55
	175-225	HR, A, N, CD	80	75	70	45
	225-275	N, CD, Q and T	65	60	50	30
	275-325	Q and T	50	45	40	25
	325-375	Q and T	35	30	30	20
	375-425	Q and T	30	15	15	10
Free Machining Alloy Steels (Resulfurized)						
4140, 4150	175-200	HR, A, N, CD	110	100	90	60
	200-250	HR, N, CD	90	90	80	50
	250-300	Q and T	65	60	55	30
	300-375	Q and T	50	45	40	25
	375-425	Q and T	40	35	30	15
Free Machining Alloy Steels (Leaded)						
41L30, 41L40, 41L47, 41L50, 43L47, 51L32, 52L100, 86L20, 86L40	150-200	HR, A, N, CD	120	115	100	65
	200-250	HR, N, CD	100	95	90	60
	250-300	Q and T	75	70	65	40
	300-375	Q and T	55	50	45	30
	375-425	Q and T	50	40	30	15

continued

Table 7-1. Recommended Cutting Speeds in Feet per Minute for Milling Plain Carbon and Alloy Steels. See Milling Column. (continued)

Material AISI and SAE Steels	Hardness HB[a]	Material Condition	Cutting Speed, fpm HSS			
			Turning	Milling	Drilling	Reaming
Alloy Steels						
4012, 4023, 4024, 4028, 4118, 4320, 4419, 4422, 4427, 4615, 4620, 4621, 4626, 4718, 4720, 4815, 4817, 4820, 5015, 5117, 5120, 6118, 8115, 8615, 8617, 8620, 8622, 8625, 8627, 8720, 8822, 94B17	125-175	HR, A, N, CD	100	100	85	55
	175-225	HR, A, N, CD	90	90	70	45
	225-275	CD, N, Q and T	70	60	55	35
	275-325	Q and T	60	50	50	30
	325-375	Q and T	50	40	35	25
	375-425	Q and T	35	25	25	15
1330, 1335, 1340, 1345, 4032, 4037, 4042, 4047, 4130, 4135, 4137, 4140, 4142, 4145, 4147, 4150, 4161, 4337, 4340, 50B44, 50B46, 50B50, 50B60, 5130, 5132, 5140, 5145, 5147, 5150, 5160, 51B60, 6150, 81B45, 8630, 8635, 8637, 8640, 8642, 8645, 8650, 8655, 8660, 8740, 9254, 9255, 9260, 9262, 94B30	175-225	HR, A, N, CD	85	75	75	50
	225-275	N, CD, Q and T	70	60	60	40
	275-325	N, Q and T	60	50	45	30
	325-375	N, Q and T	40	35	30	15
	375-425	Q and T	30	20	20	15
E51100, E52100	175-225	HR, A, CD	70	65	60	40
	225-275	N, CD, Q and T	65	60	50	30
	275-325	N, Q and T	50	40	35	25
	325-375	N, Q and T	30	30	30	20
	375-425	Q and T	20	20	20	10
Ultra High Strength Steels (Not AISI)						
AMS 6421 (98B37 Mod.), AMS 6422 (98BV40), AMS 6424, AMS 6427, AMS 6428, AMS 6430, AMS 6432, AMS 6433, AMS 6434, AMS 6436, AMS 6442, 300M, D6ac	220-300	A	65	60	50	30
	300-350	N	50	45	35	20
	350-400	N	35	20	20	10
	43-48 HRC	Q and T	25	…	…	…
	48-52 HRC	Q and T	10	…	…	…
Maraging Steels (Not AISI)						
18% Ni Grade 200, 18% Ni Grade 250, 18% Ni Grade 300, 18% Ni Grade 350	250-325	A	60	50	50	30
	50-52 HRC	Maraged	10	…	…	…
Nitriding Steels (Not AISI)						
Nitralloy 125, Nitralloy 135, Nitralloy 135 Mod., Nitralloy 225, Nitralloy 230, Nitralloy N, Nitralloy EZ, Nitrex I	200-250	A	70	60	60	40
	300-350	N, Q and T	30	25	35	20

[a] Abbreviations designate: HR, hot rolled; CD, cold drawn; A, annealed; N, normalized; Q and T, quenched and tempered; and HB, Brinell hardness number.

Speeds for turning based on a feed rate of 0.012 inch per revolution and a depth of cut of 0.125 inch.

Table 7-2. Recommended Cutting Speeds in Feet per Minute for Milling Tool Steels.
See Milling Column.

Material Tool Steels (AISI Types)	Hardness HB[a]	Material Condition	Cutting Speed, fpm HSS			
			Turning	Milling	Drilling	Reaming
Water Hardening W1, W2, W5	150-200	A	100	85	85	55
Shock Resisting S1, S2, S5, S6, S7	175-225	A	70	55	50	35
Cold Work, Oil Hardening O1, O2, O6, O7	175-225	A	70	50	45	30
Cold Work, High Carbon High Chromium D2, D3, D4, D5, D7	200-250	A	45	40	30	20
Cold Work, Air Hardening A2, A3, A8, A9, A10	200-250	A	70	50	50	35
A4, A6	200-250	A	55	45	45	30
A7	225-275	A	45	40	30	20
Hot Work, Chromium Type H10, H11, H12, H13, H14, H19	150-200	A	80	60	60	40
	200-250	A	65	50	50	30
	325-375	Q and T	50	30	30	20
	48-50 HRC	Q and T	20	…	…	…
	50-52 HRC	Q and T	10	…	…	…
	52-54 HRC	Q and T	…	…	…	…
	54-56 HRC	Q and T	…	…	…	…
Hot Work, Tungsten Type H21, H22, H23, H24, H25, H26	150-200	A	60	55	55	35
	200-250	A	50	45	40	25
Hot Work, Molybdenum Type H41, H42, H43	150-200	A	55	55	45	30
	200-250	A	45	45	35	20
Special Purpose, Low Alloy L2, L3, L6	150-200	A	75	65	60	40
Mold P2, P3, P4, P5, P6	100-150	A	90	75	75	50
P20, P21	150-200	A	80	60	60	40
High Speed Steel M1, M2, M6, M10, T1, T2, T6	200-250	A	65	50	45	30
M3-1, M4, M7, M30, M33, M34, M36, M41, M42, M43, M44, M46, M47, T5, T8	225-275	A	55	40	35	20
T15, M3-2	225-275	A	45	30	25	15

[a] Abbreviations designate: A, annealed; Q and T, quenched and tempered; HB, Brinell hardness number; and HRC, Rockwell C scale hardness number.

Speeds for turning based on a feed rate of 0.012 inch per revolution and a depth of cut of 0.125 inch.

169

Table 7-3. Recommended Cutting Speeds in Feet per Minute for Milling Stainless Steels. See Milling Column.

Material	Hardness HB[a]	Material Condition	Cutting Speed, fpm HSS			
			Turning	Milling	Drilling	Reaming
Free Machining Stainless Steels (Ferritic)						
430F, 430F Se	135-185	A	110	95	90	60
(Austenitic), 203EZ, 303, 303Se, 303MA, 303Pb, 303Cu, 303 Plus X	135-185	A	100	90	85	55
	225-275	CD	80	75	70	45
(Martensitic), 416, 416Se, 416Plus X, 420F, 420FSe, 440F, 440FSe	135-185	A	110	95	90	60
	185-240	A,CD	100	80	70	45
	275-325	Q and T	60	50	40	25
	375-425	Q and T	30	20	20	10
Stainless Steels						
(Ferritic), 405, 409, 429, 430, 434, 436, 442, 446, 502	135-185	A	90	75	65	45
(Austenitic), 201, 202, 301, 302, 304, 304L, 305, 308, 321, 347, 348	135-185	A	75	60	55	35
	225-275	CD	65	50	50	30
(Austenitic), 302B, 309, 309S, 310, 310S, 314, 316, 316L, 317, 330	135-185	A	70	50	50	30
(Martensitic), 403, 410, 420, 501	135-175	A	95	75	75	50
	175-225	A	85	65	65	45
	275-325	Q and T	55	40	40	25
	375-425	Q and T	35	25	25	15
(Martensitic), 414, 431, Greek Ascoloy	225-275	A	60	55	50	30
	275-325	Q and T	50	45	40	25
	375-425	Q and T	30	25	25	15
(Martensitic), 440A, 440B, 440C	225-275	A	55	50	45	30
	275-325	Q and T	45	40	40	25
	375-425	Q and T	30	20	20	10
(Precipitation Hardening), 15-5PH, 17-4PH, 17-7PH, AF-71, 17-14CuMo, AFC-77, AM-350, AM-355, AM-362, Custom 455, HNM, PH13-8, PH14-8Mo, PH15-7Mo, Stainless W	150-200	A	60	60	50	30
	275-325	H	50	50	45	25
	325-375	H	40	40	35	20
	375-450	H	25	25	20	10

[a] Abbreviations designate: A, annealed; CD, cold drawn: N, normalized; H, precipitation hardened; Q and T, quenched and tempered; and HB, Brinell hardness number.

Speeds for turning based on a feed rate of 0.012 inch per revolution and a depth of cut of 0.125 inch.

Table 7-4. Recommended Cutting Speeds in Feet per Minute for
Milling Ferrous Cast Metals. See Milling Column.

Material	Hard-ness HB[a]	Material Condition	Cutting Speed, fpm HSS			
			Turning	Milling	Drilling	Reaming
Gray Cast Iron						
ASTM Class 20	120-150	A	120	100	100	65
ASTM Class 25	160-200	AC	90	80	90	60
ASTM Class 30, 35, and 40	190-220	AC	80	70	80	55
ASTM Class 45 and 50	220-260	AC	60	50	60	40
ASTM Class 55 and 60	250-320	AC, HT	35	30	30	20
ASTM Type 1, 1b, 5 (Ni Resist)	100-215	AC	70	50	50	30
ASTM Type 2, 3, 6 (Ni Resist)	120-175	AC	65	40	40	25
ASTM Type 2b, 4 (Ni Resist)	150-250	AC	50	30	30	20
Malleable Iron						
(Ferritic), 32510, 35018	110-160	MHT	130	110	110	75
(Pearlitic), 40010, 43010, 45006, 45008, 48005, 50005	160-200	MHT	95	80	80	55
	200-240	MHT	75	65	70	45
(Martensitic), 53004, 60003, 60004	200-255	MHT	70	55	55	35
(Martensitic), 70002, 70003	220-260	MHT	60	50	50	30
(Martensitic), 80002	240-280	MHT	50	45	45	30
(Martensitic), 90001	250-320	MHT	30	25	25	15
Nodular (Ductile) Iron						
(Ferritic), 60-40-18, 65-45-12	140-190	A	100	75	100	65
(Ferritic-Pearlitic), 80-55-06	190-225	AC	80	60	70	45
	225-260	AC	65	50	50	30
(Pearlitic-Martensitic), 100-70-03	240-300	HT	45	40	40	25
(Martensitic), 120-90-02	270-330	HT	30	25	25	15
	330-400	HT	15	–	10	5
Cast Steels						
(Low Carbon), 1010, 1020	100-150	AC, A, N	110	100	100	65
(Medium Carbon), 1030, 1040, 1050	125-175	AC, A, N	100	95	90	60
	175-225	AC, A, N	90	80	70	45
	225-300	AC, HT	70	60	55	35
(Low Carbon Alloy), 1320, 2315, 2320, 4110, 4120, 4320, 8020, 8620	150-200	AC, A, N	90	85	75	50
	200-250	AC, A, N	80	75	65	40
	250-300	AC, HT	60	50	50	30
(Medium Carbon Alloy), 1330, 1340, 2325, 2330, 4125, 4130, 4140, 4330, 4340, 8030, 80B30, 8040, 8430, 8440, 8630, 8640, 9525, 9530, 9535	175-225	AC, A, N	80	70	70	45
	225-250	AC, A, N	70	65	60	35
	250-300	AC, HT	55	50	45	30
	300-350	AC, HT	45	30	30	20
	350-400	HT	30	…	20	10

[a] Abbreviations designate: A, annealed; AC, as cast; N, normalized; HT, heat treated; MHT, mallea-bilizing heat treatment; and HB, Brinell hardness number.

Speeds for turning based on a feed rate of 0.012 inch per revolution and a depth of cut of 0.125 inch.

Table 7-5. Recommended Cutting Speeds in Feet per Minute for Milling Light Metals.
See Milling Column.

Material Light Metals	Material Condition[a]	Cutting Speed, fpm HSS			
		Turning	Milling	Drilling	Reaming
All Wrought Aluminum Alloys	CD	600	600	400	400
	ST and A	500	500	350	350
All Aluminum Sand and Permanent Mold Casting Alloys	AC	750	750	500	500
	ST and A	600	600	350	350
All Aluminum Die Casting Alloys	AC	125	125	300	300
	ST and A	100	100	70	70
except Alloys 390.0 and 392.0	AC	80	80	125	100
	ST nd A a	60	60	45	40
All Wrought Magnesium Alloys	A, CD, ST, and A	800	800	500	500
All Cast Magnesium Alloys	A, AC, ST, and A	800	800	450	450

[a] Abbreviations designate: A, annealed; AC, as cast; CD, cold drawn; ST and A, solution treated as aged.

Calculating the Cutting Speed

The formulas for calculating the speed of the milling machine spindle and the cutter are given below for inch and for metric unit. Since the calculated speed may not be available on the machine, the closest available speed should be used. On some machines the range between speeds is large and it may be advisable to use the closest lower speed available.

$$N = \frac{12V}{\pi D} \qquad \text{(Inch units only) (7-1)}$$

$$N = \frac{1000V}{\pi D} \qquad \text{(Metric units only) (7-2)}$$

Where:

N = Spindle and milling cutter speed; rpm

V = Cutting speed; fpm, or mjmin

D = Diameter of milling cutter; in. or mm

= 3.14 (pi)

Example 7-1

A 1/2 inch (12.7 mm) diameter high-speed-steel end mill has four teeth and is to cut an O2 oil-hardening tool steel having a hardness of 200-220 HB. The cutting speed for this steel is 50 fpm (50 × .3048 = 15.2 m/min). Calculate the spindle speed, using both inch and metric units.

$$N = \frac{12V}{\pi D} = \frac{12 \times 50}{\pi \times 0.5} \qquad\qquad N = \frac{1000V}{\pi D} = \frac{1000 \times 15.2}{\pi \times 12.7}$$

$$= 382 \text{ rpm} \qquad\qquad\qquad = 381 \text{ rpm}$$

Feed Rates

Guessing a feed rate can cause the milling cutter teeth to be overloaded or drastically under loaded, each of which will have an adverse effect on the cutter. The milling machine table feed rate is expressed in terms of inches per minute (in./min), or millimeter; per minute (mm/min) on metric machines. To convert from one to the other: multiply in./min by 25.4 to obtain mm/ min; divide mm/min by 25.4 to obtain in./min. Recommended values of the basic feed rate are given in Table 7-6 for different types of milling cutters and for different materials.

Table 7-6. Feed in Inches per Tooth (fl) for Milling with High-Speed Steel Cutters

Material	Hardness, HB	End Mills							Plain or Slab Mills	Form Relieved Cutters	Face Mills and Shell End Mills	Slotting and Side Mills
		Depth of Cut, .250 in. Cutter Diam., in.			Depth of Cut, .050 in. Cutter Diam., in.							
		1/2	3/4	1 and up	1/4	1/2	3/4	1 and up				
		Feed per Tooth, inch										
Cast Steel	100-180	.001	.003	.003	.001	.002	.003	.004	.003-.008	.004	.003-.012	.002-.008
	180-240	.001	.002	.003	.001	.002	.003	.003	.003-.008	.004	.003-.010	.002-.006
	240-300	.001	.002	.002	.0005	.002	.002	.002	.002-.006	.003	.003-.008	.002-.005
Zinc Alloys (Die Castings)	...	.002	.003	.004	.001	.003	.004	.006	.003-.010	.005	.004-.015	.002-.012
Copper Alloys (Brasses & Bronzes)	100-150	.002	.004	.005	.002	.003	.005	.006	.003-.015	.004	.004-.020	.002-.010
	150-250	.002	.003	.004	.001	.003	.004	.005	.003-.015	.004	.003-.012	.002-.008
Free Cutting Brasses & Bronzes	80-100	.002	.004	.005	.002	.003	.005	.006	.003-.015	.004	.004-.015	.002-.010
Cast Aluminum Alloys—As Cast	...	.003	.004	.005	.002	.004	.005	.006	.005-.016	.006	.005-.020	.004-.012
Cast Aluminum Alloys—Hardened	...	.003	.004	.005	.002	.003	.004	.005	.004-.012	.005	.005-.020	.004-.012
Wrought Aluminum Alloys—Cold Drawn	...	.003	.004	.005	.002	.003	.004	.005	.004-.014	.005	.005-.020	.004-.012
Wrought Aluminum Alloys—Hardened	...	.002	.003	.004	.001	.002	.003	.004	.003-.012	.004	.005-.020	.004-.012
Magnesium Alloys	...	.003	.004	.005	.003	.004	.005	.007	.005-.016	.006	.008-.020	.005-.012
Ferritic Stainless Steel	135-185	.001	.002	.003	.001	.002	.003	.003	.002-.006	.004	.004-.008	.002-.007
Austenitic Stainless Steel	135-185	.001	.002	.003	.001	.002	.003	.003	.003-.007	.004	.005-.008	.002-.007
	185-275	.001	.002	.003	.001	.002	.002	.002	.003-.006	.003	.004-.006	.002-.007
Martensitic Stainless Steel	135-185	.001	.002	.002	.001	.002	.003	.003	.003-.006	.004	.004-.010	.002-.007
	185-225	.001	.002	.002	.001	.002	.002	.003	.003-.006	.004	.003-.008	.002-.007
	225-300	.005	.002	.002	.0005	.001	.002	.002	.002-.005	.003	.002-.006	.002-.005
Monel	100-160	.001	.003	.004	.001	.002	.003	.004	.002-.006	.004	.002-.008	.002-.006

continued

Table 7-6. Feed in Inches per Tooth (ft) for Milling with High-Speed Steel Cutters (continued)

Material	Hardness, HB	End Mills — Depth of Cut, .250 in. — Cutter Diam., in.			End Mills — Depth of Cut, .050 in. — Cutter Diam., in.				Plain or Slab Mills	Form Relieved Cutters	Face Mills and Shell End Mills	Slotting and Side Mills
		1/2	3/4	1 and up	1/4	1/2	3/4	1 and up	Feed per Tooth, inch			
Free Machining Plain Carbon Steels	100-185	.001	.003	.004	.001	.002	.003	.004	.003-.008	.005	.004-.012	.002-.008
Plain Carbon Steels, AISI 1006 to 1030; 1513 to 1522	100-150	.001	.003	.003	.001	.002	.003	.004	.003-.008	.004	.004-.012	.002-.008
	150-200	.001	.002	.003	.001	.002	.002	.003	.003-.008	.004	.003-.012	.002-.008
AISI 1033 to 1095; 1524 to 1566	120-180	.001	.003	.003	.001	.002	.003	.004	.003-.008	.004	.004-.012	.002-.008
	180-220	.001	.002	.003	.001	.002	.002	.003	.003-.008	.004	.003-.012	.002-.008
	220-300	.001	.002	.002	.001	.001	.002	.003	.002-.006	.003	.002-.008	.002-.006
Alloy Steels having less then 3% Carbon. Typical examples: AISI 4012, 4023, 4027, 4118, 4320, 4422, 4427, 4615, 4620, 4626, 4720, 4820, 5015, 5120, 6118, 8115, 8620, 8627, 8720, 8822, 9310, 93B17	125-175	.001	.003	.003	.001	.002	.003	.004	.003-.008	.004	.004-.012	.002-.008
	175-225	.001	.002	.003	.001	.002	.003	.003	.003-.008	.004	.003-.012	.002-.008
	225-275	.001	.002	.003	.001	.001	.002	.003	.002-.006	.003	.003-.008	.002-.006
	275-325	.001	.002	.002	.001	.001	.002	.002	.002-.005	.003	.002-.008	.002-.005
Alloy Steels have 3% Carbon or more. Typical examples: AISI 1330, 1340, 4032, 4037, 4130, 4140, 4150, 4340, 50B40, 50B60, 5130, 51B60, 6150, 81B45, 8630, 8640, 86B45, 8660, 8740, 94B30	175-225	.001	.002	.003	.001	.002	.003	.004	.003-.008	.004	.003-.012	.002-.008
	225-275	.001	.002	.003	.001	.001	.002	.003	.002-.006	.003	.003-.010	.002-.006
	275-325	.001	.002	.002	.001	.001	.002	.003	.002-.005	.003	.002-.008	.002-.005
	325-375	.001	.002	.002	.001	.001	.002	.002	.002-.004	.002	.002-.008	.002-.005
Tool Steel	150-200	.001	.002	.002	.001	.002	.003	.003	.003-.008	.004	.003-.010	.002-.006
	200-250	.001	.002	.002	.001	.002	.002	.003	.002-.006	.003	.003-.008	.002-.005
Gray Cast Iron	120-180	.001	.003	.004	.002	.003	.004	.004	.004-.012	.005	.005-.016	.002-.010
	180-225	.001	.002	.003	.001	.002	.003	.003	.003-.010	.004	.004-.012	.002-.008
	225-300	.001	.002	.002	.001	.001	.002	.002	.002-.006	.003	.002-.008	.002-.005
Ferritic Malleable Iron	110-160	.001	.003	.004	.002	.003	.004	.004	.003-.010	.005	.005-.016	.002-.010
Pearlitic-Martensitic Malleable Iron	160-200	.001	.003	.004	.001	.002	.003	.004	.003-.010	.004	.004-.012	.002-.008
	200-240	.001	.002	.003	.001	.002	.003	.003	.003-.007	.004	.003-.010	.002-.006
	240-300	.001	.002	.002	.001	.001	.002	.002	.002-.006	.003	.002-.008	.002-.005

Milling Machine Operations

Milling machines are used to perform a large variety of machining operations. In addition to those that can be classified as strictly milling operations using milling cutters, other operations, such as slotting, drilling, boring, reaming, etc., which do not utilize milling cutters and are performed on other machine tools, are often also performed on the milling machine. Although much of the work done on a milling machine involves the production of plane or contoured surfaces, large- and small-diameter holes are also frequently produced. Many of the principles pertaining to the operation of other machine tools discussed in previous chapters are used in conjunction with work commonly done on the milling machine. These include the principles of drilling, reaming, boring, and precision hole location. See Figure 7-7.

Figure 7-7 Here is an example of a bench-top mill set up and ready to go. The work is held in place by a rotary table that is sold as an accessory to this mill.
Courtesy of Sherline Products Inc.

Conventional and Climb Milling

Conventional milling is also called *up milling*. The direction of motion of the milling cutter tooth as it engages the work is opposite from the direction of the movement of the work caused by the table feed. The cutting forces resulting from this method of milling will keep the feed screw nut against the same side of the feed screw thread as when feeding the table toward the cutter without taking a cut. Thus, the table and the workpiece will never have a tendency to pull toward the cutter because of lost motion between the nut and the table feed screw.

In conventional milling a very thin chip is formed at the beginning of the cut. The thickness of the chip increases as the tooth proceeds along its path until it reaches a maximum in the position where the tooth leaves the workpiece.

Climb milling has an advantage when certain materials, such as aluminum, are milled, because it produces a much better surface finish on the workpiece than can be obtained by conventional milling. Climb milling is also called *down milling*. The milling cutter tooth and the workpiece move in the same direction. The velocity of the milling cutter tooth is faster than the velocity of the table feed, which moves the work into the cutter and thereby forms the chip. The cutting force resulting from climb milling is in the same direction as the feed. In effect, since the workpiece will be pulled into the cutter by the action of the cutting forces, the workpiece, the cutter, and the milling-machine arbor can all be seriously damaged. Climb milling, therefore, must not be used in most instances. Light profiling-type cuts can often be taken with end milling cutters using the climb milling method. The magnitude of the cutting forces is usually low, and the weight of the table is sufficient to prevent the workpiece from being pulled into the cutter. Sometimes clamping the table lightly will add an additional drag to the table so that the work will not be pulled into the cutter.

In climb or down milling, the maximum chip thickness occurs when the tooth makes the initial contact with the workpiece. As the cut continues, the chip thickness decreases, reaching a minimum where the tooth leaves the workpiece.

Setting Up the Workpiece

For most jobs done on a milling machine, setting up the workpiece is the most difficult and critical part of the work. The workpiece must not only be securely clamped, but also be held on the machine in such a position that each surface to be machined will, when finished, be accurately aligned with other surfaces on the part. Accuracy in making a setup is essential on most jobs; without it, close tolerance work cannot be done, unsatisfactory workpieces that have to be scrapped will result. Each setup must be planned in advance and then carried out with care and patience.

Planning the Setup. While each setup is unique, there are several basic types of setups. For example, large workpieces are usually placed on the top of the milling machine table (Figure 7-8) and clamped in place using strap clamps and T-bolts. The heel blocks should hold the clamps in a level position and the T-bolts should be placed as close to the workpiece as possible. See Figure 7-9 and Chapter 4 for more information on the correct and incorrect use of strap clamps.

Figure 7-8 The workpiece is aligned using a combination square.
Photo by Tom Lipton

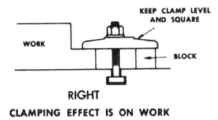

KEEP CLAMP LEVEL
AND SQUARE

WORK

BLOCK

RIGHT
CLAMPING EFFECT IS ON WORK

Figure 7-9 Here are the correct and incorrect ways to hold work in position using strap clamps and bolts.

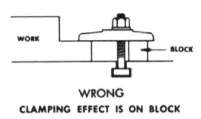

WORK

BLOCK

WRONG
CLAMPING EFFECT IS ON BLOCK

Frequently, the workpiece can be set up by clamping it in a milling machine vise as in Figure 7-10. Whenever possible, the forces generated by the milling cutters should be directed against the solid jaw of the milling machine vise rather than against the movable jaw.

*Figure 7-10
Mill vises are often used to hold work in place. The one shown here is clamped to a tilting table. Note the protractor to align the table to the proper angle.*

Before actual work in making the setup is started, the top of the milling machine table and all of the surfaces that are attached to the table must be clean and free of small chips, nicks, or burrs. Even small imperfections will prevent these surfaces from being accurately seated on the table. This condition will surely be reflected on the surfaces that are machined during the setup. As a final check before placing a part on the table, run your bare fingers over the table top and the seating surface; this procedure will detect the presence of any small chips, nicks, and burrs. Follow a similar procedure if you are placing the workpiece in a vise.

Aligning the Workpiece. Although the setup for each part is unique, usually the process becomes a matter of aligning some surface or axis on the workpiece with the spindle on the milling machine. To begin, use a dial test indicator to first make sure the mill is square and level. Follow the manufacturer's directions for setting up the tool. For some jobs, you can eyeball the alignment by adjusting the mill's table and spindle to line up with layout lines on the workpiece. But a more accurate technique is to use a dial test indicator that is attached to the tool's spindle as in Figure 7-11. Not only does an indicator help you align the work, it can also be used to find the center of a hole and check that the setup is square.

Figure 7-11 Use a dial indicator to position work accurately.
Photo by James Harvey

Settings at an angle to the spindle axis can be made by holding a protractor against a finished surface while indicating along the blade while adjusting the table. The side of the table or the table T-slots can serve as a reference surface for the protractor.

After the workpiece, vise, or angle plate has been aligned, clamp it firmly to the milling machine table. It is good practice to check the alignment again after clamping. See Figure 7-12.

Figure 7-12 An odd-shaped workpiece is clamped directly to the mill table.

Courtesy of Sherline Products Inc.

Milling Keyseats

Keyseats are slots that are cut lengthwise in shafts in which keys are held. The keys are used to transmit the driving torque that is conveyed to or from the shaft by pulleys, gears, or sprockets which are attached to the shaft. When required, keyseats are also used to align various machine elements which are sometimes attached to shafts. In addition, keys are occasionally used on machine elements other than shafts and parts that are attached to shafts.

Milling machines are usually used to machine keyseats in shafts. Keyseats are frequently cut with end milling cutters, in which case either a horizontal- or a vertical- spindle milling machine can be used. A vertical milling machine or a vertical spindle attachment is preferred for cutting key seats, because it is easier to align the cutter with respect to the workpiece, and the operation is easier for the milling-machine operator to observe.

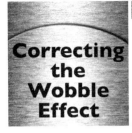

Correcting the Wobble Effect

End milling cutters are very versatile cutters that can perform a wide variety of operations. End mills are often used to cut slots. One problem that can occur when slotting with an end milling cutter is the "parallelogram" or "wobble" slot, which is a slot with sides that are parallel to each other but not perpendicular to the bottom of the slot. This condition occurs most frequently when a two-fluted end mill is used with a large helix angle. It also occurs when an excessive flute length is projecting from the spindle in which the end mill is held. The principal cause of the parallelogram slot is the deflection of the end milling cutter brought about when one flute is cutting into the material while the other flute is not cutting and is unsupported by a side of the slot. This condition is prevented by increasing the spindle speed and decreasing the feed rate so that the chip load on each tooth is reduced. Decreasing the length of the end that is projecting from the end of the spindle will improve the rigidity of the setup and thereby reduce the tendency of the end mill to deflect and to produce a parallelogram slot. The parallelogram slot can also be prevented by using a four-fluted center-cutting type of end mill, which receives better support from the sides of the slot. When the correct spindle speed and table feed rates are used, and when the setups of the work and the cutting tool are rigid, a straight slot can be cut with a two-fluted end mill.

Keyseats that are cut with ordinary four-fluted end milling cutters must extend to the end of the shaft or to the end of a shoulder in order to allow the cutter to enter the workpiece. These cutters cannot be sunk into the workpiece in the manner of a twist drill because the end teeth do not extend to the center of the cutter. Center-cutting-type end milling cutters on which the end cutting teeth extend to the center of the cutter are available. These cutters can have two, three, and four flutes, although the two-fluted center-cutting-type end mills are the most common.

Since these cutters can plunge directly into solid metal, they can be used to mill keyseats in the central portions of shafts. A type of key with a half-moon shape, called a Woodruff Key, is sometimes used. The keyseats for these keys are also half-moon shaped slots which must be cut with a special Woodruff Keyseat cutter.

The shaft in which the keyseat is to be milled can be aligned and held in a milling machine in several different ways. It can be held in a milling machine vise, or, if it is very long, in two vises. Larger shafts are sometimes clamped directly onto the table over a T-slot, which helps to align the shaft on the milling machine. The shaft can also be clamped in a V-block, or in a matched pair of V-blocks, as shown in Figures 7-13 and 7-14. The clamps holding the shaft in place are not seen in these illustrations. V-blocks can be aligned on the milling-machine table by placing them against a slot block, as shown in Figure 7-13 or on V-blocks with a key in their base which fits into the T-slots of the table (see Figure 7-14).

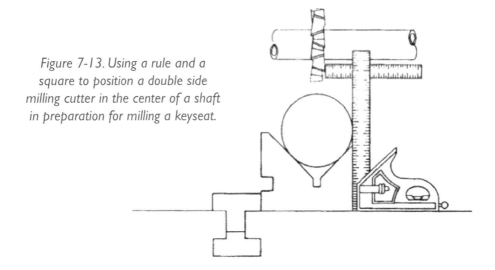

Figure 7-13. Using a rule and a square to position a double side milling cutter in the center of a shaft in preparation for milling a keyseat.

Aligning the Shaft. Before the keyseat can be cut, the milling cutter must be aligned with the center of the shaft. This can be done by first making a layout of the keyseat on the shaft and adjusting the table in the transverse direction until the cutter is within the layout lines. Frequently, however, the keyseat is cut without a layout in order to save time. In this case one of the procedures described in the following paragraphs can be used to center the cutter with respect to the shaft.

A square and a steel rule, as shown in Figure 7-13, can be used to align the cutter with the center of the shaft. The square is held against the side of the shaft, and the distance from the blade of the square to the side of the milling cutter is measured with a rule. The milling cutter should, of course, not be rotating. The table of the milling machine is adjusted until the measurement made by the rule is equal to one-half of the difference obtained when the width of the diameter of the end milling cutter is subtracted from the diameter of the shaft.

The other procedures depend on making accurate movements of the table which are obtained by reading the micrometer dial of the transverse feed screw. One method is to hold the blade of a square against the side of the workpiece and to move the table until the milling cutter, which must not be rotating, touches the blade as shown at A in Figure 7-14. The arrows in Figure 7-14 indicate the direction that the table must be moved in order to reach the positions shown. It is helpful to place a thin paper feeler between the cutter and the square to gage the contact between the blade of the square and the milling cutter. The paper feeler should slip; however, a drag should be perceptible when it is pulled. The cutter and the blade of the square are then the thickness of the paper feeler apart. The distance that the table should be moved in order to center the cutter over the shaft should then be equal to one-half of the difference between the diameter of the shaft and the thickness of the diameter of an end milling cutter minus the thickness of the paper feeler.

Before the table is moved to the center position, as determined by the reading of the micrometer dial, it should be moved beyond this position as shown at B in Figure 7-14. The table is then moved to the center position illustrated at C in Figure 7-14. This procedure must be used to eliminate the error that can result from the lost motion between the feed screw and the feed screw nut. It should be noted that the micrometer dial must be read when the table has reached positions

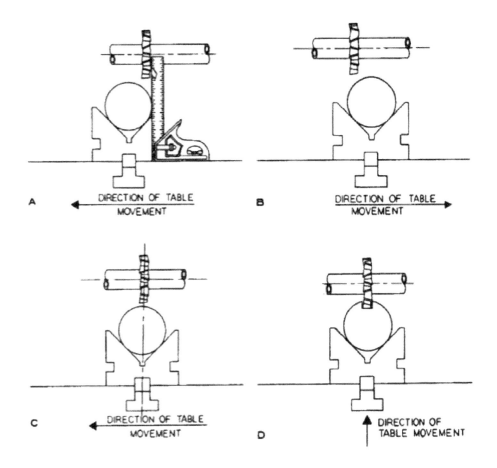

A

DIRECTION OF TABLE
MOVEMENT

B

DIRECTION OF TABLE
MOVEMENT

C

DIRECTION OF TABLE
MOVEMENT

D

DIRECTION OF
TABLE MOVEMENT

Figure 7-14. Procedure for centering a milling cutter in preparation for milling a keyseat.
A. Touching up against blade of square using a feeler. B. Compensating for lost motion in feed screw by
moving beyond center. C. Positioning the table to center the cutter. D. Cutting the keyseat.

A and C in order for it to move the exact distance required to align the cutter and the shaft. Also, observe that the table is moved in the same direction to reach both position A and position C. The table is then moved vertically to obtain the required depth of cut, and the keyseat is cut to the required length as shown at D, Figure 7-14.

The third procedure can be used when it is possible to touch the side of the shaft with the milling cutter, as shown in 7-15. In this illustration a four-fluted end milling cutter held in a vertical spindle is used. The milling cutter should be rotating while a long strip of paper is held between the cutter and the shaft with one hand. Simultaneously the other hand is used to turn the transverse feed hand wheel in order to move the shaft toward the cutter slowly and carefully. When the teeth of the cutter just graze the paper without cutting into it, the table movement is stopped. The table is then centered by moving it a distance equal to the sum of the radius of the shaft, the radius of the end milling cutter, and the thickness of the paper feeler. For example, if the diameter of the shaft is 1.500 inches, the diameter of the end milling cutter is .500 inch, and the thickness of the paper feeler is.003 inch, the table movement required to center the cutter would be 1.500/2 +.500/2 + .003 = 1.003 inches.

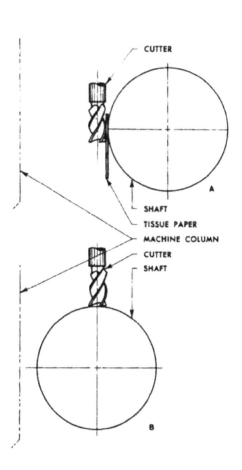

Figure 7-15.
Procedure for centering a milling cutter in
preparation for milling a keyseat by
touching up against the side
of the workpiece.

Courtesy of Cincinnati Milacron

Tip from a Pro

The Rotary Table

Text and photos reprinted by permission of
J. Randolph Bulgin
and The Home Shop Machinist
(July/August 2009).

I have received more requests for information about rotary tables than any other facet of machining, so here goes nothing. Hopefully, this article will have one of two effects. It will either give you some information or leave you so hopelessly confused that you will seek out the information from another source.

The rotary table/circular milling attachment/circular dividing table/that round thingy with the crank on the side of it – whatever you call it where you come from—is likely the second most used work holding attachment for the vertical milling machine to be found in the small shop. The milling vise, of course, being the first. They come in many shapes and sizes. I have seen rotary tables, my preferred name for the attachment, as small as 3" in diameter and as large as 40 feet. The most common in our smaller shops, and what we are going to talk about here, is more likely to be between 6" and 12" in diameter. *Photo 1* is a picture of the unit I use most in my shop. It is an 8", Japanese-made rotary table that has served me well for over thirty years and still is as good as the day I bought it. The plug setting on top of it is a centering plug made from the shank of a used up No. 3 Morse drill bit. But you aren't interested in my rotary table, so let's talk about the one you either already own or you are going to buy and what you can do with it after you buy it. If you already own one then you are one step ahead of the game.

Photo 1 My rotary table. The plug setting on top is a centering plug made from the Morse shank of a drill bit.

When selecting a rotary table here are some of the characteristics you need to consider:

- Size. It should be as large as you can conveniently use with your milling machine. The lower the profile the better, as sometimes you will be limited in the height of the job you can accommodate on your machine. A 24" rotary table can conceivably be set up on a mill-drill with an 18" table, but it wouldn't be of much use to you.
- Having a rotary table with the ability to be set up in either the horizontal or vertical position on the milling machine table is a good thing. Here again you will have to consider the size of your milling machine.
- It should have positive clamps for clamping the table in a fixed rotary position. These tables are, like all manual mechanical movements in the shop, subject to backlash.
- You will find it useful if your rotary table has a Morse taper in the center.
- It is not necessary for there to be some way of disengaging the worm so that the table may be rotated by hand, but it is useful.
- And although it isn't exactly on the subject, a DRO is really useful when doing some operations with a rotary table. At the very least, you should have a couple of long travel dial indicators to keep track of exactly where your work is relative to the tool.

Other features are available. There are rotary tables with X/Y positioning capability, tilting rotary tables, power driven rotary tables, and maybe even singing rotary tables, but many times, in my experience, I have found that the more features built into a device, the more difficult it becomes to use for its intended purpose.

SETUP

Aside from all the usual considerations of setting up a job on the milling machine—good hold down bolts and clamps, clean surfaces between the device and the machine table, making sure that there will be no interference between cutter and clamps, etc.—there are two important considerations. Your milling machine spindle needs to know where the center of your rotary table is, and your rotary table needs to know where the center of your part is. These locations, although equally important, have nothing to do with each other. In

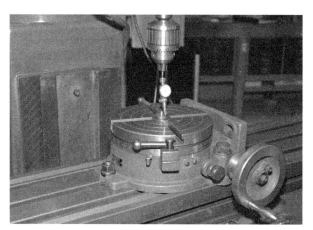

Photo 2 Locating the center of the rotary table relative to the axis of the milling machine spindle.

Photo 2, we are indicating the position of the rotary table relative to the spindle axis of the milling machine. The indicator is mounted in the spindle and is rotated about the centering plug of the rotary table. The centering operation is accomplished by movement of the X and Y axes of the milling machine.

Photo 3 shows the alternate operation. The indicator is mounted so as to bear on the work piece and the rotary table is turned—by hand if you have the means of disengaging the worm gear—to center the part. The centering operation in this situation is accomplished by moving the part from side to side atop the rotary table. If you cannot disengage the worm you will have to crank the table through several revolutions until the part is centered.

It is important that you understand that both these centering operations must be done for most rotary table work, but they are separate operations and are completely independent of each other.

Now let's make some chips. The part I am using here as an example *(Detail 1)* is just that, an example. It can hopefully demonstrate a few of the capabilities of the rotary table. *Photo 4* shows the material set up to begin. The milling machine knows where the rotary table is, and the rotary table knows where the material is. Notice the wooden parallels holding the part clear of the table surface. Wood is a good choice here because we can cut into it without damaging the cutting tool. If you use parallel bars, and sometimes you will want to do that, place them as carefully as you can,

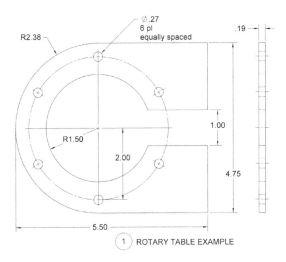

1 ROTARY TABLE EXAMPLE

Detail 1

Photo 3 Centering the workpiece on the rotary table.

Photo 4 Setting up to make the sample part. The spindle is aligned with the "X" on the work piece.

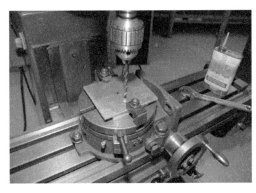

Photo 5 Drilling a hole to begin milling the center bore.

considering your tool path. Many times you will need to reposition your clamps and/or your parallel bars as you machine the part. When this is required, try to always leave at least one clamp, preferably two, in place. In this example we will only be using two clamps. At this beginning point, we have the DRO set to zero in both axes, and the rotary table is set to zero degrees and clamped into positions.

Photo 6 Machining the bore.

Move the saddle of the milling machine, the Y-axis, a distance equal to one half the diameter of the bore minus half the diameter of the cutting tool. I used a drill to provide a starting place for the milling cutter, but a center cutting end mill would work here. Penetrate the part all the way through and then rotate the table through 360° to machine the bore. Good machining practice here would suggest that you rough the part out leaving .010" or so and then climb mill to produce the finish cut (*Photos 5, 6, 7, and 8*).

Photo 7 Finishing the bore by climb milling the last few thousandths of an inch.

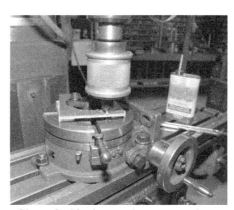

Photo 9 Machining the outer radius

Photo 8 Finished bore and slot.

Photo 10 Finishing the outer radius.

Of course, you are being careful to keep all machine movements clamped except for the one being used. This includes the rotary table, the X-axis, and the Y-axis of the milling machine.

Up until now we have not really been concerned with the angular position of the rotary table. The hole is a full circle and it didn't matter where we started and/or finished. But for the 1" slot we must be oriented a little more accurately. Move the Y-axis of the milling machine back to zero and rotate the table to 0°. Now move the Y-axis one-half the width of the slot minus one-half the diameter of the cutter in either direction. Lock the table against rotation; lock the Y-axis against movement; and traverse the X-axis to make the cut. Now unlock the Y-axis, and move to the opposite side of the slot and repeat. Again, you should leave .010" or so for a finish cut.

Now let us go to the outside of the part: The radius that is tangent to the straight lines of the sides. First we must move the hold down clamps to the inside of the part as shown in **Photos 9 and 10**. Remember to not to loosen all clamps at one time. You must always leave at least one in place; otherwise, you will move the part and have to start over. After repositioning the clamps, move the Y-axis of the machine to one-half the outside dimension plus one-half the cutter diameter. Then traverse the table and the X-axis to clear the end of the part and machine the first side. Machine to the center (or X-zero), clamp the X-axis, unclamp the table, rotate 180°, clamp the table, unclamp the X-axis, and machine to the end of the part.

The last operation on this part is drilling the bolt pattern. Move the Y-axis of the milling machine to a point equal to the radius of the bolt pattern, keeping the X-axis at zero. Ensure that the rotary table is locked at 0°, and drill the first hole. From then on it is just a matter of cranking the table around 60° at a time and drilling the remainder of the holes. See **Photo 11** and the finished part in **Photo 12**.

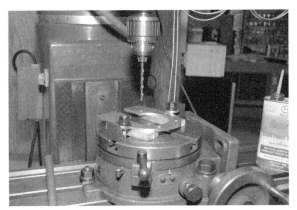

Photo 11 Drilling the bolt hole pattern.

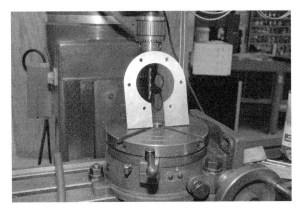

Photo 12 The finished sample part.

189

Using the rotary table set up in the vertical position is a little different in some ways but just the same in others. ***Photos 13, 14 and 15*** show steps in setting up the table for use in this position. In **Photo 13** we see the table being squared with the milling machine table. If a higher degree of accuracy is required than may be obtained by this method, a dial indicator must be used. If you have a good machinist's square this is quite accurate. ***Photos 16, 17 and 18*** are pictures I made for a former project, but the sequence provides a good demonstration of what may be done with the rotary table in the vertical position.

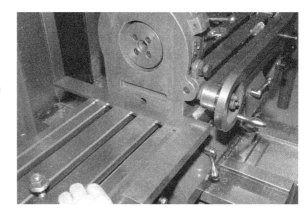

Photo 13 Squaring the rotary table to the milling machine table.

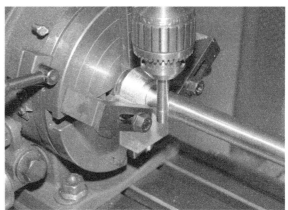

Photo 14 Locating the center of the rotary table relative to the spindle axis.

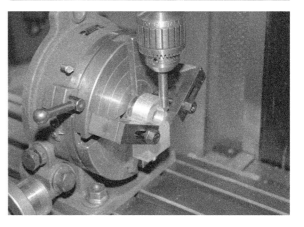

Photo 15 Locating the spindle axis relative to the X-axis of the part.

*Photo 16, 17 and 18
Machining in the vertical position.*

It seems that we always include in our discussions of small or hobby machine shops some sort of reference to CNC and what that has done to our trade. CNC machines have made it possible to machine contours and configurations never before possible. Or at least not possible without requiring a huge amount of effort and innovation on the part of the machinist. That imagination and ability to be innovative is still present by the way, it is just now freed to open new horizons. The rotary table, the tracer duplicating systems on both lathes and milling machines, rotary milling heads, and many other attachments and accessories have disappeared from the mainline manufacturing shops, but that is a good thing for those of us who enjoy making parts our way. It makes them more readily available to us.

Glossary

A

abrasive: A manufactured or natural very hard, tough material capable of wearing away another material softer than itself. Used for grinding and improving a surface by friction, as in polishing, buffing, cleaning. Sandpaper, steel shot, glass beads, and steel wool are also examples of abrasives. Abrasives can vary in hardness from diamond (hardest), silicon, carbides, emery, feldspar, rouge (softest).

abrasive dressing stick: An abrasive stick, usually hand held in a holder, used to dress the face of smaller-size grinding wheels. Because it also shears the grains of the grinding wheel, it is often used for roughing or preshaping, prior to final dressing with a diamond dresser. See also grinding wheel dresser.

abrasive wheel: See grinding wheel.

Acme thread: A screw thread, either outside (male) or inside (female), that has an included angle of 29°. Acme threads have a section that is between the square and V and are used extensively for feed screws, to cause movement, or to generate power transmission through the rotation of the male thread within the female thread. This thread is easier to machine than a square thread and was developed to carry heavy loads without causing excessive bursting pressure in the hole or nut.

additives, cutting oils: A substance used to increase the effectiveness of basic mineral cutting oils. See also soluble oils.

adhesive: Any substance, organic or inorganic, that is capable of fastening or bonding by means of surface attachment. The bond durability depends on the strength of the adhesive to the base material or substrate (adhesion) and the strength within the adhesive (cohesion). Surface preparation is very important, and the presence of oil, grease, mold-release agents, or even a fingerprint can destroy a good bond. Surface preparation such as chemical etching or mechanical roughening may be needed to improve joint strength on some materials.

adjustable angle plate: Also know as sine angle plates, this tool is precision made and has provisions for setting them with micrometer or rectangular gage blocks. A sine bar is frequently used when an adjustable angle plate is not available.

adjustable tap: A tool for cutting the thread of an internal screw, made with slots in the shank (tap body) for holding separate blades or chasers that can adjusted to control the tap's cutting edge.

adjustable tap wrench: A tool containing two opposing handles and a V-shaped opening in the center. One of the handles is adjustable making it possible to hold the square ends of taps and reamers of various sizes in the V-shaped opening. *See also* tap wrench.

AISI/SAE steel designations: A four digit number system used to designate the chemical composition of carbon and alloy steels. This system describes plain carbon and low to medium alloy content, used primarily in machine parts. The first two numerals designate either plain carbon or the alloy grouping and quantity, and the last two numerals give the mean carbon content in hundredths of a percent. The first digit for carbon steels is 1. Plain carbon steels are designated 10xx. Plain carbon steels containing 0.75% carbon are designated 1075, etc.

Additionally letters are added between the first and last pairs of digits to designate special qualities and additives as follows:

B: indicates the presence of boron in amounts of 0.0005-0.003% for enhanced depth-hardening.

L: indicates the presence of lead in amounts of 0.15-0.35% for enhanced machinability.

M: indicates merchant quality steel, for alloy steels .

E: indicates electric furnace steel, for alloy steels.

H: indicates hardenability requirements, for alloy steels.

Allen screw: Specialized set screws, socket-head screws, and machine bolts containing a hollow head with a hexagonal shaped hole that is used for tightening and loosening the screw and made to accept, and be adjusted by, an Allen wrench.

Allen wrench: Also known as hex wrenches or hollow-setscrew wrench. Available in assorted sizes, these L-shaped hand wrenches are hexagonal-shaped bar stock of hardened steel, used to fit into hollow-headed Allen screws (safety set-screws), socket-head screws and fasteners.

allowance: (1) The amount of acceptable clearance, or the desired difference in dimensions, between mating parts; the minimum working clearance or maximum interference prescribed (intentionally permitted) difference used to achieve various classes of fits between different parts. Allowances may be positive (sliding fit) or negative (force fit); (2) in screw threads, the prescribed difference between the design (maximum material) size and the basic size. It is numerically equal to the absolute value of the ISO term *fundamental deviation.* Not to be confused with tolerance.

alloy: A substance formed by melting any combination of metals together, or the mixture of one or more metals with non-metallic elements, as in cast iron or carbon steels. One metal is usually in much larger proportion than the others. An alloying element is used to improve the properties of an alloy. The major classes of alloys are ferrous and non-ferrous, depending on whether or not iron is a component.

angle of drill point: The commercial standard for angle of drill point 118° included angle, 12° to 15° lip clearance, angle of chisel point 125° to 135° is best suited for drills engaged in all average classes of work.

angle gage blocks: Precision tools made from hardened tool steel that is ground and lapped to precision tolerances, used to accurately measure and inspect angles. Like rectangular gage blocks, these angle blocks can be wrung together in various combinations. Complete sets contain 16 blocks measuring 4 in. on the base with 5/8 in. thickness that can be used to accurately measure 356,400 angles ranging from 1 second up to 99 degrees.

annealing: A heat-treatment process that decreases the hardness and brittleness of metal (and other crystalline materials) by relieving internal stresses by recrystallization. The work is heated and held at a definite high temperature and then allowed to cool at a relatively slow rate. In addition to removing hardness, the process may also result in grain refinement and improved mechanical properties.

anvil: A block of cast steel or iron, with an upper horizontal surface of hardened steel on which metals are forged or hammered.

arbor: (1) A machining term for a shaft, spindle, or bar used holding, supporting, and driving cutting tools for milling, grinding, drill press, bandsawing, or other machining operations. An arbor frequently has a taper shank fitting the spindle of a machine tool. An arbor is not the same as a mandrel (a workholding device), and the terms should not be used interchangeably; (2) a foundry term describing reinforcements in sand molds made from metal.

automatic center punch: A metal punch that contains an internal spring-controlled hammer. When sufficient pressure is applied to the handle of the punch by hand (as no hammer blow is used)

B

back rest: In turning and grinding, a tool used to support slender work.

back taper: The slight decrease in diameter from point to back in the body of the drill.

backlash: (1) Wear in gear, screw mechanism, or other moving parts that can result in lost motion, slippage, or *end play* causing vibration; (2) a possibly dangerous reaction involving the sudden release of potential energy of a body in motion when the body stops, usually causing the body to quickly reverse directions. *Down milling* on milling machines require a backlash eliminator.

ball end mill: A shank-type end milling tool used for milling pockets, fillets, and slots with rounded bottoms. These mills have end-cutting teeth that can be used for plunge milling or longitudinal milling.

ball-peen hammer: The most commonly used machinist's hammer. The head is made of tool steel having one face that is slightly convex (called the flat face), and the other a ball face or peen. The flat face is used for driving center punches, chisels, and for various other general purposes, the peen (or rounded) end is used for purposes such as flattening rivet heads. Available in many sizes from 1 oz (28.35 grams) to 3 lb. (1.36 kg), the 16 oz. (453.6 grams) size is generally preferred for general shop work.

base metal: (1) The primary metallic element of an alloy to which other elements are added (e.g., copper in brass); (2) the original core metal to which coating of plating or cladding is applied.

basic size: (1) The theoretical or nominal standard size from which all variations are made; (2) that size from which the limits of size are derived by the application of allowances and tolerances; (3) the size to which limits of deviation are assigned. The basic size is the same for both members of a fit.

bastard file: A file between coarse and second cut having approximately 30 teeth per inch; used for fast removal of work material. By comparison, a coarse or rough-cut file has approximately 20 teeth per inch and a second-cut file has approximately 40 teeth per inch.

bell-mouthed slot or groove: A condition where the ends of the slot or groove become gradually wider than the center.

bench vise: The most commonly used general-purpose vise. The bench vise is bolted securely to a workbench and usually has a swivel base.

bevel: (1) Any flat surface not at right angle (90°) to the rest of the piece; (2) another name for a bevel square.

bevel gear: A gear having beveled teeth used for transmitting rotary motion at an angle. Bevel gears connect shafts which are not parallel to each other, and whose center lines meet each other at right angles.

bevel protractor: An instrument that contains the features of a bevel and a protractor that can be used and adapted to all classes of work where angles are to be accurately laid out. Depending on the need for degrees of accuracy, bevel protractors can be constructed with or without a vernier dial.

bevel square: A tool, similar in appearance to a try square or steel square, but having a blade hinged to the stock which allows it to be moved and set to any desired angle in its own plane.

billet: A solid block of material usually semifinished and in the form of cylinder or rectangular prism that has been cast or hot worked by forging, rolling, or extrusion preparatory to some finishing process.

blanking: Cutting or punching out predesigned flat shapes from metal sheets. This is nearly always the first operation in producing the finished article. Blanking may be combined with other operations in one tool, all the work being performed by a single stroke of the press.

blind hole: A hole that has been drilled into, but not completely through, the workpiece.

blind hole tapping: A process of cutting threads to a specified depth. This is usually done in a blind hole. Often called bottom-hole tapping.

body: That portion of a drill extending from the shank or neck to the outer corners of the cutting lips.

body diameter clearance: A drill term use to describe that portion of the land that has been cut away so it will not rub against the wall of the hole.

bore: The width or diameter of a hole.

boring: The process of internal finishing that includes enlarging, and finishing off a previously drilled (generally the hole is drilled undersize by 1/16 in to 1/8 in) or cored hole to accurate size (truing), using a single-point, cutting tool called a boring tool or boring bar which travels along the inside of the work as it revolves. Boring can be accomplished on lathes and with boring heads on drill presses and milling machines, or on specially built boring machines. The workpiece must be clamped securely to the machine table and boring must be done at low rpm with automatic feed.

bottoming tap: A tap used for cutting internal threads or "tapping" blind holes (hole that go only part way into the work) to the bottom of the hole.

Brinell hardness test: A measure of the relative hardness of the smooth surface of a metallic material, obtained by measuring the resistance it offers to the depth of indentation of a standard 10 mm (0.394 in) hard steel or carbide ball at a standard load of 3,000 kg (6600 lb) for a period of 15 seconds for steel and 30 seconds for nonferrous metals. The greater the distance of penetration, the softer the work, and the higher the Brinell harness number (BHN).

buffing: The process of smoothing a metal objects by holding and pressing them against a fabric disk or belt imbedded with loosely applied buffing compound, running at high speed. The buffing compound consists of fine abrasive particles held in a composition of wax or similar binding materials and often formed into hand-held sticks or cylinders. See also polishing.

burr: A sharp, thin, usually jagged sliver of metal left on a workpiece as the tool from a machine or punch operation exits the cut.

butt joint: A joint between two components or plates, lying in the same plane. The ends of the plates to be joined abut squarely against each other. By comparison, a lap joint has the plates to be joined overlapping each other.

C

caliper: An instrument used for testing the work where the accuracy of a micrometer is not required and for measuring the dimensions of work pieces, especially internal and external diameters of cylindrical pieces. The caliper consists of two pieces of steel that are curved and are hinged together with a tight joint at one end, the distance between the points representing the measurement taken. In general, calipers are used to measure distances between or over surfaces, or for comparing distances with standards, such as those on a graduated rule. Their size is measured by the greatest distance they can be opened between the two points. Never use calipers on work that is moving or revolving.

carbon steels: An alloy of iron having carbon as its chief alloying element. Mild (low-carbon) steels contain 0.02-0.25% carbon; medium steels contain 0.25-0.7% carbon; and, high-carbon grades contain 0.7-1.5% carbon. Also known by the names, "plain carbon steel," "ordinary steel," and "straight carbon steel."

cast iron: A general term used to describe a wide range of ferrous alloys; a saturated solution of carbon in iron. The amount of carbon can vary from 1.7% to about 6%, depending upon the amount of silicon (usually about 1% to 3%), manganese, phosphorus and sulfur present in the solution. Cast iron weighs about 0.26 lb/in^3; tensile strength 15,000 - 30,000 lb/in^2 according to grade. Cast iron is also used to describe remelted pig iron. Cast iron is usually machined dry.

C-clamps: A general-purpose clamp having a cast or drop-forged frame in the shape of a "C" and a screw having a "V" type thread, but larger sizes are usually made with square threads.

center drill: A combination tool for drilling and countersinking a shallow hole in one operation, used in predrilling operations such as accurately locating and guiding the drill bit to a hole center. The center drill has a cutting edge angle of 60°. Also used to make bearing surfaces for lathe centers that are properly shaped. Also known as combination drill and countersink, center drills are available in several sizes.

center punch: A hand punch with a good, solid shank to withstand hammer blows and a taper toward the point to allow the mark to be seen clearly. The shank is knurled to provide good finger grip, and the top end is slightly chamfered to prevent the edge from becoming burred from constant hammer blows. Used for marking the center of a point or position, usually for starting a drill. The point of the center punch should be a sharper angle than the point of the drill, to insure the drill starting true. The point of a center punch is usually ground to a conical shape of 90°.

chamfer: (1) *noun*: A beveled surface on a workpiece used to eliminate an otherwise sharp edge that can become damaged; (2) *verb*: The act of cutting or "softening" the edge of a workpiece at an angle of less than 90°; (3) in thread design, the conical surface at the starting end of a thread; (4) in tapping, the tapering of the threads at the front end of each land of a tap by cutting away and relieving the crest of the first few teeth to distribute the cutting action over several teeth.

chatter: A problem affecting the accuracy and finish of metals, caused by rapid vibration of the tool away from the work or *vice versa* and identified by the little ridges, grooves, or lines (chatter marks) appearing at consistent intervals on the surface of the workpiece. The spacing of the chatter marks depends upon the frequency of vibration. Chatter is often self-sustaining and can be prevented or reduced by correcting the problem which can be caused by excessive speed, too light a feed or rate of table travel, loose spindle, poor choice of cutting material, unbalanced workpiece, etc.

chipbreaker: A tool feature such as a groove that prevents a continuous chip from growing to such a length that it ruins the work, or becomes a nuisance or safety hazard.

chips: Small pieces of material removed from a workpiece by cutting tools, or by abrasion.

chisel: See cold chisel.

chisel edge: A drill term used to describe the edge at the ends of the web that connects the cutting lips.

chuck: A device that is mounted on a machine tool spindle, used to hold a rotating cutting tool or workpiece. There are many different types of chucks that can be actuated manually, hydraulically, or pneumatically. The universal chuck has three or four jaws which are controlled to move together, making it easy to center round stock with a fair degree of accuracy. The independent chuck is made to hold work between four jaws, which are adjusted independently and allow irregular shaped pieces, as well as regular shapes, to be positioned for lathe operations. The Jacobs chuck is mounted on a taper shank to fit in either the tailstock or headstock of the lathe. It is used primarily in the tailstock for center drilling and boring.

chuck key: A device used for adjusting chuck jaws.

clearance: Generally, space allowed to prevent interference. However, the term "clearance" should not be used in specifications without indicating clearly just what it means.

clearance diameter: A drill term used to describe the diameter over the cutaway portion of the drill lands.

clearance fit: (1) A fit having limits of size so specified that a clearance always results when mating parts are assembled; (2) the relationship between assembled parts when clearance occurs under all tolerance conditions; (3) in thread design, a fit having limits of size so prescribed that a clearance always results when mating parts are assembled at their maximum material condition.

climb milling: A form of milling in which the rotating teeth of the milling cutter "climbs" or comes down on the work in the same direction as the feed at the point of contact. Differs from the standard milling operation in that, instead of the job passing under the cutter against the rotation of the teeth, it is fed in the same direction as the path the teeth take. The teeth cut downward rather than upward. Advantages of this method is that play between feed-screw and nut is eliminated, and does not tend to lift the job, as in standard milling; increased tool life since chips pile up behind the cutter, improved surface finish since chips are less likely to be carried by the tooth, easier chip ejection since chips fall behind the cutter, and decreased power requirements since a higher rake angle can be used on the cutting tool. The cutting machine should be equipped with a backlash eliminator attachment. See also milling.

clutch: A mechanical device used to link an engine and transmission. The clutch disconnects the motor from the transmission, allowing gears to be changed. When the clutch is reengaged the engine and transmission resume contact causing both to turn together at the same new speed.

CNC: Abbreviation for computer numerical control.

cold chisel: A hand tool made in several shapes and are identified by the shape of their cutting edges (the angle can vary from 50° to 75°, with 60° for most general work). They are used for chipping flat surfaces, for cutting cold metal, removal of rivet heads, and cutting nuts and bolts which are rusted fast. Available in various sizes from 1/4 in. to 1 in. diameter, and 4 in. to 8 in. length.

collet: (1) A cone-shaped split-sleeve bushing with a hole for holding circular or rodlike tools and workpieces (i.e. drill, reamer, or tap) by their outside diameter during grinding and machining. The split-sleeve allows the hole to be reduced in size. A collet generally provides greater gripping force and precision than a chuck utilizing jaws; (2) a small, self-centering chuck used on a lathe.

combination squares: A tool containing square, spirit level, and protractor all mounted on an adjustable blade that can be made to slide along the head and be clamped at any desired place. The sliding blade contains a concave center groove which travels a guide in the head of the square and can be pulled out and used simply as a rule. The spirit level in the head can be used as a simple level and can be used to square a piece with a surface and at the same time ascertain whether one or the other is level or plumb. Also, the head of the combination square may be removed from the blade and replaced with an auxiliary *center head,* an instrument used to find the center of cylindrical pieces such as a shaft.

complementary angles: Any two angles whose sum is 90°, whether they are adjacent or not, are called *complementary*.

compound rest: A lathe fixture that slides and is attached to the carriage cross-slide; used for holding the tool post.

compound slide: An essential part of the lathe that supports the tool post and swivels the cutting tool on the horizontal plane by adjusting its base which is graduated 90° each way from the center.

conventional milling: Also known as up-milling. Milling in which the workpiece is rotating in the opposite direction of the table feed at the point of contact. Chips are cut starting with minimal thickness at initial engagement of cutter teeth with the work, and increase to a maximum thickness at the end of engagement.

corner joint: A joint formed between two components located approximately at 90° to each other.

corrosion: The gradual electrochemical or chemical change to metals or alloys caused by reaction with the environment. This reaction is accelerated by the presence of moist acids or alkaline environments, and corrosion products often take the form of metallic oxides or sulfides.

counterbore: A cutting tool having a small end called a *pilot* used to guide the drill bit into the hole and keep it centered. The pilot should be oiled prior to counterboring. See also counterboring.

counterboring: A drilling process of enlarging previously drilled holes. This process can be used to enlarge one end of a hole to provide countersinking or a flush seat for a screw (such as a fillister-head cap screw) or nut, or to obtain better surface finish and tolerances when machining cast, forged, pressed, or extruded materials. Counterboring is called spot-facing if the depth is shallow.

countersink: A boring tool used to enlarge the entrance of a hole with a conical depression or beveled edge to receive the head of a screw, usually a flat headed screw that can sit flush with the surface of a workpiece. These bits have different cutting edges depending on the screws being used. Countersinks may be used in a hand drill, bit brace, or drill press.

countersinking: The process of using a countersink. The size of the countersink hole depends upon the head of the screw.

critical speed: A rotation speed of shafts and other bodies at which vibrations due to unbalanced forces reach a maximum.

cross feed: (1) In lathe work, movement of the cutting tool across the end of the workpiece; (2) in milling or surface grinding, movement of the table toward or away from the column.

cross-slide: A fixture attached to the lathe carriage, that can be moved in and out, used for holding the compound rest.

cutoff: A lathe operation for parting off a workpiece that is mounted in a lathe chuck. Also used to describe parting off using a cutoff wheel.

cutoff tool: Also called a *parting tool*. A lathe tool or blade made from relatively thin steel used for parting off stock, cutting grooves, and cutting to a shoulder. The sides of a cutoff tool have top-to-bottom clearance tapers and should never be ground.

cutting fluids: Natural or synthetic liquids used to cool the cutting tool and workpiece, wash away chips, lubricate the bearing formed between the chip and lip of the cutting tool. Used to enable the cutting tool to produce a good finish, extend the cutting tool's life, improve dimensional accuracy, and protect the finished product from rust and corrosion. Cutting fluids are used on cutting tools such as drills, taps, dies, reamers, milling cutters, lathe cutting tools, and power saws.

cutting speed: The linear or circumferential or peripheral distance traveled in one minute by a point on the tool or the workpiece in the principal direction of cutting; expressed in feet or meters per minute.

D

deformation: (1) The altering of the shape, flow, or elasticity of a material without rupture; disfigurement, as the elongation of a test piece under tension test.

degree: (1) A division or interval marked on a scale, generally a difference in temperature (as °F). Or direction (as °angle); (2) condition in terms of some unit or in relation to some standard.

depth gage: A measuring device or tool which contains a narrow rule or rod with a sliding head (also known as sliding stock or base) set at right angles to the rule and with a means of clamping the slide to lock the reading. Used for determining the depth of holes, recesses, shoulders, slots, keyways, grooves, etc.

depth of cut (DOC): The distance a cutting tool is advanced into the revolving workpiece, measured at right angle to the piece.

design size: The design size is the basic size with allowance applied, from which the limits of size are derived by the application of tolerances. Where there is no allowance, the design size is the same as the basic size.

dial indicator: An instrument containing a graduated dial, used by inspectors, toolmakers, and machinists in setup and inspection work. Available with inch or metric graduation dials that show the amount of error in size or alignment of a part. Dial indicators can be used on snap gages, depth gages and for other purposes. Depending on the degree of accuracy required, the dials may be graduated in thousandths of an inch up to 50 millionths of an inch or, in hundredths of a millimeter to two-thousandths of a millimeter.

die: (1) A tool used to cut outside (external) threads; (2) a tool or device used to produce a specific shape or design in a material such as a metal either by stamping, die casting, extrusion, forging, or by compacting powdered metal; (3) a mold used in die casting; (4) a device through which alloys are drawn to make wire, filaments, etc.

dog: (1) A general name used to describe any projecting piece which strikes and moves some other part; (2) a device, such as a lathe dog, used for clamping a workpiece so that it can be revolved by faceplate of a lathe. The bent tail dog is used for driving round, square, hexagonal or other regular work and the clamp dog for rectangular work. Some better grades of dogs contain safety screws, and are balanced to eliminate vibrations.

double cut, file: The double cut file has a multiplicity of small pointed teeth inclining toward the point of the file arranged in two series of diagonal rows that cross each other. For general work, the angle of the first series of rows is from 40 to 45° and of the second from 70 to 80°. For double cut finishing files the first series has an angle of about 30° and the second, from 80 to 87°. The second, or upcut, is almost always deeper than the first or overcut. Double cut files are usually employed, under heavier pressure, for fast metal removal and where a rougher finish is permissible.

drill diameter: A drill term used to describe the diameter over the margins of the drill measured at the point.

drill press: A machine tool used for small, light work and using smaller drills. This small press consists of an adjustable table for holding the work and a frame in which are mounted a vertical revolving spindle to carry the drill which is connected to a handle, and a motor or other mechanism to drive the drill at varying rates of speed. This is the simplest drill press, it can be mounted on a floor stand or used on a bench, in which case it may be called a bench drill.

E

elasticity: The ability of a material to assume its original form, volume, size, and shape after a force causing deformation or distortion is removed.

embossing: A process of forming a pattern, shallow indentations, or raised designs on the surface of metal objects.

end milling cutter: Also called end mills. A milling cutter with cutting edges on both its circumference (face) and end (periphery). The teeth on the circumference may be straight or helical. End mills can be solid with teeth and shank in one piece, or the shankless shell-type for mounting on an arbor. Solid end mills typically have an integral straight or taper shank for mounting. The straight shank contains a flat surface for securing it in the mill end holder with a set screw.

F

face, file: The widest cutting surface or surfaces that are used for filing.

faceplate: The disk or plate that screws on the nose of a machine spindle or lathe and drives or carries work to be turned or bored. Sometimes the table of a vertical boring machine is called a faceplate.

facing: The process of making a cut across the end of a workpiece. When facing on the lathe, the cutting tool should be set exactly in line with the workpiece center, and the cutting edge of the cutting tool should be set at an angle of 80° to the face.

feed: Also called feed rate. The rate at which work is advanced relative to the position of the cutting tool. An important factor in determining the rate of metal removal and overall machine efficiency. Quality of finish, effects on the machine, and safety are important considerations in selecting the proper feed rate. Usually expressed in inches per minute (ipm).

file: A hand tool used for cutting and smoothing. A file consists of a blade or body with a tang that fits into a wooden handle. The blade is hardened and tempered, and teeth of a suitable kind are cut into the blade's faces and edges, although the edges of some files are smooth and uncut (termed "safe"). Files are classified according to their shape or cross-section and according to the pitch or spacing of their teeth and the nature of the cut. The cross-section may be quadrangular, circular, triangular, or some special shape.

flute (1) Shop name for a groove, applied to taps, reamers, drills and other tools; 2) a drill term used to describe the helical or straight grooves cut or formed in the body of the drill to provide cutting lips, to permit removal of chips, and to allow cutting fluid to reach the cutting lips.

flutes: (1) Longitudinal recesses in cylindrical parts; (2) the helical grooves or straight, longitudinal channels cut or formed in a tool such as a tap to create cutting edges on the thread profile and to provide chip spaces and cutting fluid passages; (3) Surface imperfections found in formed or drawn parts.

flycutting: A specialized milling operation using one or more rotating single-point tool bits or cutters mounted on a fly cutter arbor which is attached to the spindle of a milling machine.

G

gage: (1) The thickness or diameter of a material; (2) to measure accurately; (3) a tool, instrument, or device for establishing a particular dimension of an object, for determining whether or not a workpiece is within specified limits. Except for the surface gage which is *adjustable to a fixed range*, the great variety of gages are fixed.

gage blocks, rectangular: Precision tools made of hardened, ground, and stabilized tool steel blocks with surfaces that are flat, parallel, and finished to an accuracy within 0.000008 in., used as the standard of precision measurement.

galvanizing: (1) Adding an atmospheric-protective finish to a metal by coating it with a less oxidizable metal; (2) to coat iron or steel with a thin corrosion-resistant coating of zinc.

gib: A guide located alongside a sliding member to take up wear, or to ensure proper fit.

gray cast iron: A soft cast iron usually containing from 1.7 to 4.5% carbon and graphite (the allotropic form of carbon), and from 1 to 3 % silicon. It is tough with low tensile strength; it breaks with a coarse grained dark or grayish fracture. The excess carbon is in the form of graphite flakes and these flakes impart to the material the dark-colored fracture which gives it its name. Gray cast iron may easily be cast into any desirable form and it may also be machined readily.

grinding: A machining operation involving the surface attrition of a material from a workpiece by rubbing with an abrasive material imbedded in a powered wheel that is mounted in a grinder or grinding machine. Grinding is used to shape or improve the surface of a workpiece.

grinding wheel: A cutting tool made from natural or artificial abrasive grains of differing size and properties that are mixed with a suitable cement or "bond," and compressed into a wheel.

gullet: The space between the cutter teeth that allows for the exiting of chips during a cutting operation.

gun metal: A series of bronzes formerly used to make cannons. An alloy of copper containing 85-88% copper, 5-10% tin, 2-5% zinc, and sometimes small amounts of antimony, nickel, lead, iron, and/or aluminum.

H

hacksaw, hand: A general work saw that consists of a handle, metal "bow" frame, and a narrow saw blade made of steel and hardened to cut metal. The frame can be of the fixed type made to take only one length of blade, or adjustable to accept blades from 8 to14 in. in length.

half round file: A file having one flat face and one curved or semi-circular face, and is always double-cut on the flat face. The round face may be double- or single-cut. Used for curved or con-cave surfaces. Made in various lengths.

hand punch: A long steel tool designed for a variety of work. The tool is held in the hand and struck with a hammer while the other end is held against the work. Hand punches are usually made of tool steel and are knurled or octagonal shaped to provide good finger grip.

hardening: A general term for any process used to increase the resistance to breaking, bending, cutting, or grinding of metal by suitable treatment, usually involving heating and cooling. Heating at even temperature above the critical range gets the grain structure in the steel into the proper state. Cooling is the quenching of the steel in some medium such as water, brine, caustic solution, or oil in order to preserve the structure caused by heating.

hardness: The most well-known property of solid materials. Although difficult to define accurately, it is measured by determining the resistance to external forces such as wear, abrasion, cutting, scratching, or plastic deformation, usually by penetration, or the effect upon the rebound of a weight.

headstock: That component of the lathe that consists of the permanently fastened housing located at the left end of the bed. It contains the motor drive system and spindle that holds and turns the workpiece.

heat treatment: A process involving a combination of heating and cooling temperature cycles ap-plied to a metal or alloy in the solid state to obtain desired conditions by changing their physical properties.

height gage: A precision instrument for measuring above a given surface.

hermaphrodite calipers: A two-legged steel instrument containing one divider-type leg with a sharp point (which may be adjustable) and one bent caliper leg and are used to scribe center lines on a round bar or shaft, or to scribe lines parallel to the edge on a surface of a workpiece. Their size is measured by the greatest distance they can be opened between the two points.

high carbon steels: Steels containing from about 50% to 1.50% carbon. Tool steels are high carbon steels and contain sufficient carbon to allow them to be hardened and tempered.

high-speed steel (HSS): Also known as high-speed tool steel. So named because they are designed primarily for efficient removal of metal faster than ordinary steel, because high speed steel retains its hardness (does not lose temper) at high temperatures. Due to this property, such tools may operate satisfactorily at speeds which cause cutting edges to reach red heat.

hole saw: A metal cylinder with teeth around its circumference and a twist drill bolted to its center and which is used as a guide. This drill is used to drill large diameter holes in relatively thin material.

I

impact resistance: The amount of force or energy required to fracture a standard-size sample in an impact test such as the Izod or Charpy test. The resistance is expressed in terms of the number of foot-pounds or meter-kilograms of force required to break the specimen with a single blow. Sometimes called impact energy or impact value.

incomplete thread: A thread design term used to describe a threaded profile having either crests or roots or both, not fully formed, resulting from their intersection with the cylindrical or end surface of the work or the vanish cone. It may occur at either end of the thread.

indexing: (1) On turret-type machines, rotating the turret to make other tools available; (2) in dividing head operations, moving a workpiece angularly and sequentially so that equally spaced divisions can be machined.

internal thread: A thread on a cylindrical or conical internal surface.

iron: Pure iron (also known as ferrite) is a relatively soft metallic element of crystalline structure. CAS number: 7439-89-6. An extremely important metal; the only metal that can be tempered; used principally in steels and other alloys.

J

jig: A work-holding device that places the workpiece in proper position and holds it securely. The jig can contain guides and hardened steel bushings for aligning, guiding and supporting various cutting tools for drilling, reaming, tapping, etc.

K

kerf: The width of a cut produced during a cutting process, as the slot or passageway cut by a saw.

keyseat: An axially located rectangular groove in a shaft or hub.

keyway: A groove, usually in a shaft and the piece which is to be fastened to it, in which a key is driven. The keyway is usually square or rectangular, but can be round or other shaped.

knurling: A lathe process using a knurling tool to impress a rough pattern design on the peripheral surface of a work blank, generally used to provide a nonslip surface for gripping, handling, or turning cylindrical pieces by hand or to improve appearance of the workpiece.

L

lamination: Metal defects involving the separation of two or more layers, sometimes found in wrought, rolled, or forged metals.

land: (1) A drill term used to describe the peripheral portion of the drill body between adjacent flutes; (2) a gear term used to describe the top land is the top surface of a tooth, and the bottom land is the surface of the gear between the fillets of adjacent teeth.

lathe: The most basic of all metalworking machine tools. A turning machine capable of producing cylindrical and conical parts by spinning the circumference of the workpiece against the cutting edge of a single-point cutting tool that is mounted on a carriage that may be fed manually or automatically along the work while machining. The cutting tool may be made of high-speed steel, cast alloys, cemented carbide, ceramics, or natural or artificial diamond. Also used for boring holes more accurately and/or larger than those produced by drilling or reaming.

lathe center: A lathe accessory with a 60° cone point that is incorporated into a lathe headstock or tailstock and used to support the workpiece.

lathe tool: Any cutting instrument held in the tool post of a lathe.

layout: The transfer of information containing lines and geometrical shapes from a working drawing to a metal surface of a workpiece.

left-hand thread: A thread is a left-hand thread if, when viewed axially, it winds in a counterclockwise and receding direction. All left-hand taps are stamped "L" or "LH." Right-hand taps are unmarked.

live center: A lathe term use to describe the headstock center that turns with the spindle or the tailstock center that turns on ball bearings.

longitudinal feed: In lathe work, the principal direction or movement of the cutting tool along the workpiece, parallel to the bed of the lathe.

low carbon steel: Steel without enough carbon (content usually below 0.03%) to allow it to harden to any great extent when heated to a specific temperature and quenched in water, oil, or brine.

M

machinability: A term used to describe the ease with which an engineering material can be machined, worked, or cut. It is based on various factors including rate of metal removal, tool life, surface roughness, and power consumption. Most common metals are rated in terms of machinability using B1112 steel at 100%, as the comparison baseline. By comparison, carbon and chromium steels are rated from about 45 to 60, while aluminum is rated from 300 to 2,000. Generally, as the machinability rating is increased, the cutting speed is also increased.

machine reaming: A drill press operation used to make light finishing cuts to a predrilled and/or bored, undersize holes with a tool called a *fluted machine reamer*. The feed for machine reaming should be faster than used for drilling and a cutting fluid should always be used. The speed for machine reaming should be slower, perhaps by one-half to two-thirds that of drilling.

malleability: A measure of a metal's ability to endure hammering or rolling without breaking, fracturing, or returning to its original shape.

margin: (1) A drill term used to describe the cylindrical portion of the land, which is not cut away, to provide clearance; (2) in riveting, a term used to describe the distance from the edge of the plate to the center line of the nearest row of rivets.

medium carbon steels: Steel with a carbon content of about .20% to .60%. Where resistance to wear is required, medium carbon steels may be heat treated increase its strength or case hardened by cyaniding or by pack hardening, or quenched directly from the furnace.

metallurgy: The science that deals with metals theory and the processing of metals, from their recovery as ores to their purification, alloying, and fabrication into usable industrial products.

mild steel: Carbon steel containing a maximum of about 0.25% carbon.

milling: A machining operation in which metal is removed by bringing the workpiece into contact with a horizontal or vertically mounted rotating, multiple-tip cutting tool called a mill or milling cutter.

milling arbor: A machine shaft that is inserted in the milling machine spindle that both drives and holds a rotating milling cutter.

milling cutter: Also known as a *mill*. A rotating tool containing one or multiple teeth which engages the workpiece and removes material as the workpiece moves past the cutter. Milling cutters can be termed *horizontal* with teeth (cut or inserted) on the periphery or *vertical* (usually held in a vertical spindle) with teeth on both the end and periphery.

N

neck: A drill term used to describe the section of reduced diameter between the body and the shank of a drill.

nitrites: Rust inhibiting chemical agents added to cutting fluids.

noble metal: A term applied generally to the gold, platinum, and palladium families of the periodic table, and are generally considered to be gold, silver, platinum, palladium, iridium, rhenium, mercury, ruthenium, and osmium.

nominal size: The designation used for the purpose of general identification.

O

obtuse angle: If the inclination of the arms of an angle are more than a right angle (90° but less than a straight line 180°also known as a straight angle since it corresponds to a rotation through half a full circle), the angle is called *obtuse*.

obtuse triangle: A triangle having one angle larger than 90°. Both obtuse- and acute-angled triangles are known under the common name of oblique-angled triangles. The sum of the three angles in every triangle is 180°.

outside calipers: A two-legged steel instrument with the hardened steel ends of each bent inward. Their size is measured by the greatest distance they can be opened between the two points. Used to measure widths and thickness of workpieces and the outside diameters of circular objects.

overall length: A drill term used to describe the length from the extreme end of the shank to the outer corners of the cutting lips. It does not include the conical shank end often used on straight shank drills, nor does it include the conical cutting point used on both straight and taper shank drills.

overheating: The exposure of a metal or alloy to such a high temperature that it acquires a coarse-grained structure causing its physical properties to be impaired.

oxidation: The original meaning meant simply combination with oxygen. The term in its broadest sense means a chemical reaction whereby electrons are transferred.

P

parallel clamp: Also known as a toolmakers' clamp. A clamp consisting of two pointed jaws with a screw passing through the center of each jaw and another screw at the flat or opposite end of the jaws. Used to hold parts together or to hold small parts in the pointed end by tightening the screw at the flat end of the jaws which should be kept parallel to the workpiece surfaces.

parting: An operation that uses a parting tool performed on a lathe or screw machine to remove or separate a finished part from a length of chuck- or collet-held stock..

patina: The ornamental and/or corrosion-resistant green coating that slowly forms on the surface of copper and copper alloys, such as bronzes, and other metals, that are exposed to the atmosphere or from suitable chemical treatment.

perforating: The piercing or punching of identical multiple holes arranged in a regular pattern.

personal protection equipment (PPE): Safety equipment designed to protect parts or all of the body from workplace hazards. Such protective equipment includes chemical resistant clothing, gloves, respirators, and eye protection.

pig iron: The basic raw material for cast iron and steel. A compound of about 93% pure iron ore with about 3.5% carbon, and small amounts of silicon, sulfur, phosphorus, and manganese that are obtained from smelting iron ores in a blast furnace and run into gridiron-shaped molds in the open air and slowly cooled either by spraying or running through a tank of water. Commercial pig iron is produced when iron ore is charged in a blast furnace, mixed with limestone as a flux, and melted down with either charcoal, coke or anthracite coal as fuel.

pinion: When two gears mesh together, the one with the smaller number of teeth is called the pinion, and the other is the gear. An exception occurs in worm gearing, where the gear with the smaller number of teeth is called the worm.

plain milling: A machining operation that produces flat surfaces with a plain milling cutter mounted on a horizontal milling machine arbor.

plain milling cutter: A broad category of cylindrical cutters with straight or helical teeth having a helix angle generally up to 60° on the circumference or periphery only, and available in a wide variety of widths and diameters. Plan milling cutters can be sharpened on a cutter and tool grinder by grinding the land on the periphery.

point: (1) Drill term used to describe the cutting end of a drill made up of the ends of the lands, the web, and the lips. In form, it resembles a cone, but departs from a true cone to furnish clearance behind the cutting lips; (2) a file term used to describe the front end of a file; the end opposite the tang.

polishing: A process of flexible grinding, used to smooth the surface of a workpiece by rubbing with abrasive particles imbedded in, or attached to, a soft cushioned matrix or flexible backing in the form of a wheel or belt which is moving or rotating. When the abrasive material is applied loosely, the process is called buffing.

powder coating: A process using powdered thermosetting resins such as acrylics, epoxies, and polyesters that are applied metal objects by electrostatic spraying techniques. Reportedly, this process minimizes the pollution problems encountered with solvent-based sprayed coatings.

power: In mechanics, the product of force times distance divided by time equals power; it measures the performance of a given amount of work in a given time. It is the rate of doing work and as such is expressed in foot-pounds per minute, foot-pounds per second, kilogram-meters per second, etc. The metric SI unit is the watt, which is one joule per second.

protractor: An instrument for setting off, reading, or measuring angles. In its simplest form it is a graduated semi-circle, marked out to 180° along the circumference. At the center of the protractor's straight base is a starting point such as a hole or notch which acts as a reference, the angle being read in degrees from the circumference.

punch: (1) A small hand tool with a straight or tapered end that is struck with a hammer, used for various purposes including punching holes in sheet metal, layout marking, and to transfer the location of a hole.

Q

quality control: The maintenance of a specific level of quality during the manufacturing process to ensure a consistently good product. Techniques include physical sampling and statistical quality control.

quenching: The operation of rapid cooling hot metal in a suitable medium such as liquids (water, brine, oil), gasses (air) or solids. An essential part of the tempering process, especially for steels and alloys.

quick-change toolholder: A lathe toolholder made to hold four or more different cutting tools, containing a patented locking system that enable the operator to change tools fast and accurately.

R

rake: The angle between the flute on a drill and the workpiece. An angle less than 90° is called a *positive rake*, and a rake angle more than 90° is a *negative rake*. A slightly negative rake is preferred for drilling softer materials such as brass or plastics.

rasp cut: A file tooth arrangement that is formed individually by raising a series of individual rounded teeth, usually not connected, from the surface of the file blank with a sharp narrow, punch-like cutting tool. The rasp file is used with a relatively heavy pressure on soft substances for fast removal of material.

ratio of gearing: Ratio of the numbers of teeth on mating gears. Ordinarily the ratio is found by dividing the number of teeth on the larger gear by the number of teeth on the smaller gear or pinion. For example, if the ratio is 2 or "2 to 1," this usually means that the smaller gear or pinion makes two revolutions to one revolution of the larger mating gear.

reamers: Multi-edged, fluted, rotary cutting tools used to bring a hole to accurate size and to produce a good surface finish on existing cylindrical or tapering holes, whether a cast or cored hole, or one made by a drill or boring bar. There are two main types of reamer, one of which is parallel for making straight holes and the other tapered which make tapered holes.

recess: A groove set below the normal surface of a workpiece.

reference cut: A machining term for the initial cut that is used to set the base for subsequent cutting measurements.

right triangle: A triangle having one angle that is a right or 90° angle.

right-hand thread: A thread is a right-hand thread if, when viewed axially, it winds in a clockwise and receding direction. A thread is considered to be right-hand unless specifically indicated. In other words, right-hand taps are unmarked while left-hand taps are stamped "L" or "LH."

Rockwell hardness test: A measure of relative hardness of the surface of a material based on the indentation made by a 1/16, 1/8, or 1/4 in. standard steel ball penetrator, or a conical diamond cone, called a brale indenter, with an apex angle of 120°. The steel balls are used on soft, non-ferrous metals and read on the "B" scale. The diamond point used on harder materials and read on the "C" scale (used particularly for steel and titanium). The hardness is indicated on a dial gage graduated in the Rockwell-B (RB) and Rockwell-C (RC) hardness scales. The results are reported by using numbers to denote the pressure in kilograms, and letters scale to denote the ball or diamond producing a given indentation. The harder the material the higher Rockwell number will be.

roughness (R): Describes the finely-spaced micro-geometric irregularities of the surface texture, usually including those irregularities which result from the inherent action of the production process. These are considered to include traverse feed marks and other irregularities within the limits of the roughness sampling length.

round file: A file that is generally single cut. When parallel they are called *parallel round*; tapered round files are described as *rat-tailed*. Used for enlarging holes, filing fillets, and concave radii.

S

SAE: Abbreviation for Society of Automotive Engineers. The initials of this organization are used in its tests and specifications for motor oils, fuels, and steels.

safe edge: An edge of a file that is made smooth or uncut, so that it will not injure that portion or surface of the workplace with which it may come in contact during filing.

scrapers: A hand tool for removing a very small amount of metal and used to correct the irregularities of a machined work surface so that the finished surface is a plane surface. Scrapers may be classed as flat, hook (left or right hand), half round, triangular or three cornered, two-handled, and bearing.

screw pitch: The distance from the center of one screw thread to the center of the next. In screws with a single thread, the pitch is the same as the lead but not otherwise.

screw-pitch gage: Also known as a thread gage. A tool for determining the number of internal or external threads per inch or *pitch* of a screw or nut. It consists of a holder containing a series of thin steel or stainless steel blades or "leaves," somewhat like a pocket knife. The blades are notched to create a profile and are used for comparing and measuring pitch of inside threads or inside holes and nuts as well as external nuts.

scriber: A tool with long, slender, sharp points of hardended steel used to scratch or mark lines on metal. Some scriber have a single point while others have a straight point and a point bent at 90° on the other. The bent point is used to mark lines where the straight end cannot reach such as the inside of cylindrical objects. Used for measuring or layout work.

serial taps: Made for progressive cutting internal threads in very tough metals, these taps are usually made in sets of three. The top (handle) end of the shank contains identifying marks such as one, two, or three circumferential rings to identify them as part of a set. This is important because they resemble other common tap types, but differ in both major and pitch diameter. Each tap starting with No.1 (marked with a single circle) is used in succession to start-, rough-, and final-cut the thread to its correct size.

shank: A drill term used to describe the part of the drill by which it is held and driven.

shear: (1) The effect of external forces acting so as to cause adjacent section of a member to slip past each other; (2) an inclination between two cutting edges; (3) a tool for cutting metal and other material by the closing motion of sharp, closely adjoining edges; (4) the ratio between a stress applied laterally to a material and the strain resulting from this force.

sheet metal: A general term used to describe metal products that are thinner than plates, thicker than foil, and wider than strips.

side-milling cutters: A term describing several types of milling tools that are similar to plain milling cutters but with cutting teeth on the periphery of the cutter, and depending on the type, may have teeth on one or both sides. The teeth may be straight, staggered, or helical. These cutters are used for side-milling, slotting, grooving, and *straddle milling*.

silicon steel: Steel alloys nominally containing 0.40-4.50% silicon.

single cut: A file tooth arrangement where the file teeth are composed of single unbroken rows of parallel teeth formed by a single series of cuts. Refers to the tooth-forming cut found on the faces and edges of various kinds of files. A single cut file has one series of rows of cuts running across the file face at an angle varying from 45 to 85 degrees with the axis of the file.

slitting saw: Also called metal-slitting cutters. A thin, saw-like plain milling cutter having fine pitch teeth on the periphery only and sides that are slightly tapered toward the center or hole (creating a dish-effect) to provide side relief and prevent binding. Used for common cutoff processes and for cutting narrow slots.

smooth cut: An American pattern file and rasp cut that is smoother than second cut.

snap gage: A fixed gage with inside measuring surfaces for calipering outside diameters, lengths, thicknesses of parts. Similar to and often called a caliper gage.

soldering: The process of joining metals by employing a nonferrous metal or metallic alloy filler whose melting point is lower than that of the base metal and in all cases below 800°F/427°C. The filler is applied as a thin layer, in a molten state, and drawn into the space between them by capillary action, and allowed to cool.

solvent: A chemical liquid, capable of dissolving another substance. An industrial term generally used to describe organic solvents.

square: See steel square.

square file: A file that is double cut on each face and may be parallel square or tapered for the last third of its length. Used for enlarging square holes and filing square corners, keyways, splines, and slots.

stainless steel: High-alloy steels contain relatively large amounts of chromium and possessing high strength and superior corrosion and oxidation resistance when compared to the carbon and conventional low-alloy steels. Most stainless steels contain at least 10% chromium and few contain more than 30% chromium or less than 50% iron. However, in the United States the stainless steel classification includes those steels containing as little as 4% chromium.

standard: Any established measure of extent, quantity, or quality of values.

steel: An alloy of iron and carbon and other elements including chromium, silicon, phosphorus, sulfur, manganese, aluminum, vanadium, and nickel. The terms *carbon steel* or *plain carbon steel* are used to describe steel that contains no other alloying element other than carbon.

steel rule: Rules made of tempered steel and containing graduations cut with great accuracy. The rules are made in various lengths and a great variety of graduations.

steel square: A precision measuring instrument that consists of a thick stock or *beam* and a thin blade set exactly (or as near as possible) at 90° of one another. The stock and edges are hardened and accurately ground to insure straightness and parallelism.

stress: The force applied per unit area that tends to deform a body. The load may be static (constant) or dynamic (increasing at a uniform rate). In either case it induces a strain in the material that results in rupture if the deforming force exceeds its strength. There are three kinds of stress: tensile, compressive, and shear.

stud: Also called a stud bolt. A headless bolt that is threaded at both ends. One threaded end may be permanently installed in a fixture to receive a removable part, such as a cover, and a fastening device such as a nut is affixed to the other threaded end.

surface: (1) The boundary which separates one object from another object, substance or space; (2) Mathematically, a two-dimensional entity, a figure having length and width but no thickness.

surface gage: An adjustable type gage used for many purposes including testing the accuracy or parallelism of planed surfaces, and for gaging the height between a flat surface and some point on the work. The gage can also be used for the scribing of lines at a given height from some or all faces of the work.

T

tailstock: Also referred to as dead-center. A movable fixture opposite the headstock on a lathe. The tailstock is movable and can be clamped along the bed of the lathe. It contains a spindle that does not turn but is adjustable (it can be moved in and out) and is used to support one end of a workpiece. The tapered hole of the tailstock also accepts tools for drilling operations including drill chucks, tapered shank drills, and reamers.

tailstock set over: The amount that the tailstock or dead-center on a lathe is offset from the headstock during the turning of tapers.

tang: (1) A file term use to describe the narrow, tapered or pointed end of a file that fits into a wooden handle. While the blade is hard and brittle, the tang is tempered to be soft and tough as it would otherwise be easily broken where the handle meets the blade; (2) a drill term used to describe the flattened end of a taper shank, intended to fit into a drill press spindle or in a drill sleeve.

tap: A tool for cutting internal threads in metal. A short length of cylindrical, hardened and tempered tool steel with a straight or slightly tapered thread at one end and a square shank on the other. The threaded portion has flutes cut in it parallel with the axis, which form the cutting edges and allow space for the shavings. Taps are obtainable in three forms: The taper tap, the plug tap, and the bottoming tap. Taps are used for cutting internal or female threads.

tap wrench: Sometimes called a T-handle wrench or a T-handle tap wrench. A hand tool used for holding and turning small taps and hand reamers. The T-handle tap wrench has two jaws inserted in an adjustable chuck and fits the square end of the tap. The handle with tap attached can be turned gradually into the hole and cuts threads of an interior surface.

telescoping gage: A T-shaped gage used for measuring inside diameters and the widths of slots and grooves. The gage is equipped with a plunger located at 90° to the knurled handle. The tool is inserted into a part with the plunger retracted; the plunger is released, the knurled nut on the handle is tightened to lock the plunger, and the gage is removed from the part. The distance across the top of the "T" which includes the plunger is measured with a micrometer.

tempering: Also called *drawing*. A heating and cooling process used to relieve some of the internal stress and brittleness caused by the hardening operation, and to increase toughness.

tolerance: The total amount by which a specific dimension or surfaces of machine parts is permitted to vary. The tolerance is the difference between the maximum (upper) and minimum (lower) limits of and specified dimension.

tool bits: Small pieces of hardened and sharpened material which are held, welded, or cemented in the tool holder and do the actual cutting in the lathe.

tool holder, lathe: A steel device with a rectangular body which is clamped into the tool post of a lathe, designed to rigidly and securely hold small pieces of tool steel or the working end of a cutting tool bit in a desired position. The cutting tool bit can be removed from the tool holder for sharpening or replacement without disturbing the holder. Used for the various lathe operations: turning, cutting off, boring, threading, and knurling.

tool post: That part of the lathe compound rest used to mount the tool holder.

tool steel: Any high carbon steel that is suitable for blanking and forming tools. Tool steel contains a sufficient carbon content (usually .60-1.50) that will allow hardening if heated above a certain temperature and rapidly cooled.

T-slot milling cutter: A special milling machine end-mill tool used to make T-slots such as those found on the tables of milling machines, shapers, drill presses, and other machine tools.

try square: A term used to describe any square used to test or "try" the accuracy of work. A square designed for this sole purpose would not necessarily require graduations on the blade. However, many try squares are manufactured with graduated blades that are not hardened and held in a stock containing a special nut and bolt arrangement that allows the tool to be taken apart so the worn or "out of square" blade and/or stock can be re-ground or lapped.

turning: An operation performed on a lathe using a single-point cutting tool that removes metal as it moves longitudinally along the workpiece diameter which is revolving. Turning is used to machine many different straight or tapered cylindrical shapes.

U

Unified Screw Thread: The basic standard for fastening types of screw threads in the United States, Great Britain, and Canada. This standard was agreed upon in 1948 to attain screw thread interchangeability among these nations. In relation to previous American practice, Unified threads have substantially the same thread form and are mechanically interchangeable with the former American National Screw Threads of the same diameter and pitch. Also known as Unified Form Thread.

V

valley: A term related to the measurement of surface texture, the point of maximum depth on that portion of a profile that lies below the centerline and between two intersections of the profile with the centerline.

V-block: A workpiece-holding tool made of steel and containing a 90° V-shaped groove, used to support cylindrical work on which a layout is being made, or which is being machined or inspected. The standard V-blocks are usually made in matched pairs to exact dimensions and are often fitted with a clamp which aids in holding the work in the groove.

vernier: A scale that has been added to a measuring device, used to make very fine measurements.

vise: A workpiece-holding device that can be mounted on a bench for bench work, or used as an accessory on various machines to hold work between jaws that can be manually adjusted by a screw mechanism attached to a handle, or sometimes with a toggle or lever, or can be adjusted hydraulically. The jaws can be fixed or changeable.

W

ways: The precision-machined guiding or bearing surfaces (tracks) on which moving parts slide, as a carriage or tailstock of a lathe plane or milling machine.

web: A drill term used to describe the central portion of the body that joins the end of the lands. The thin, extreme end of the web forms the chisel edge at the cutting end on a two-flute drill.

web thickness: A drill term used to describe the thickness of the web at the point unless another specific location is indicated.

Woodruff key: A remountable machinery part having the shape of a half circle, which, when assembled into key-seats, provides a positive means for transmitting torque between the shaft and hub.

Woodruff keyseat cutter: Also called keyseat cutter. A milling machine end-mill or arbor-type cutter used for cutting keyseats for standard Woodruff keys.

wrought iron: A low-carbon content steel containing a considerable amount (1-4%) of slag fibers (iron silicates) entrained in a ferrite matrix. A commercial, virtually pure form of iron made from pig iron containing about 0.035% carbon. Wrought iron is tough, malleable, and has good ductility, weldability, and corrosion resistance. Also, it is relatively soft, lacking the strength of most steels, and is expensive to produce.

X

X-axis: The horizontal axis on the Cartesian coordinate system, or coordinate plane.

Y

yield strength: The stress at which a material exhibits a specified deviation from proportionality of stress and strain. The deviation is expressed in terms of strain. An offset of 0.2% is generally used for many metals.

Z

zero-defect: A term used to describe improvements in the manufacturing process or operation whose goal is to produce products with no defects.

zinc coating: A sacrificial, corrosion-resistant coatings used on steels. The coating can be applied by hot-dipping, electroplating, etc.

Index